Basics of
Inorganic Chemistry

:: Author ::

Dr. Darshan V. Chaudhary

PUBLISHED BY

Hemchandracharya International Publishing House
H.Q. At & Po. Chaveli., Ta- Chansma,
Dist- Patan, North Gujarat, India, Asia.
www.iphouseindia.com

First Publication: 9th NOVEMBER, 2015

Copyright: Author

(c) **Dr. Darshan V. Chaudhary**

ISBN:- 978-1-51747-710-3

Price: Rs.800/- INDIA

  $ 10 OUTSIDE INDIA

PUBLISHED BY

Hemchandracharya International Publishing House
H.Q. At & Po. Chaveli., Ta- Chansma,
Dist- Patan, North Gujarat, India, Asia.
www.iphouseindia.com

PREFACE

Where has the Basic Inorganic Chemistry come from? Throughout the antiquity of humanoid competition, societies have thrashed to make intelligence of the world around them. Through the division of science we call Chemistry we have increased an understanding of the chemical calculations, atomic structures, introduction of Metallurgy, periodic classification, s & p block elements which makes up our world and of the interactions between chemical properties on which it depends.Important advances in our understanding of the nature of inorganic chemistry of different elememts and their action were made in the late 18th and 19th centuries, seeding the explosive expansion from the 1850s and 60s onward to the present billion dollar inorganic chemistry based industries.

While preparing for the examination, students should not restrictthemselves, only to the questions/problems given in the self-evaluation. They must be prepared to answer the questions and problems from the entire text.Learning objectivesmay create an awareness to understand each andevery chapter. Sufficient reference books are suggested so as to enable the students to acquire more information's about the concepts of inorganic chemistry.

In present book, chapter 1 introduces the different types of Chemical Calculations in atoms, molecules and elements. Chapter 2 describes the General Introduction to Metallurgy. Chapter 3 elaborates different Atomic Structures in Chemistryon, while Chapter 4 emphasis PeriodicClassification in different groups, Chapter 5 discusses s-Block, and the last Chapter 6 elaborates scope, the p-Block elements in Chemistry.

The contents of the book will be useful to the students of Chemistry, Biotechnology, Industrial Chemists, Pharmaceutical science and technology, Inorganic Chemistry, Public health sciences etc.

I express my heartfelt thanks to Dr. PranavSrivastav, Professor, Chemistry Department, Gujarat University, Ahmedabad, India, for his constant guidance across my research Work and without the same platform I would not be able to compile this book. I am very thankful to my other colleague contributor Mr. EdvinPithawala for critical evaluation of each chapter across the book. I would like to express my gratitude to my family members especially to my parents for their love affection and care and last but not the least to my beloved wife*Reema* for her everlasting love, motivation and sacrifice for the time taken in compiling this book.

I am grateful to publisher for their concern, efforts and encouragement, especially for their excellent cooperation in the task of preparing and publishing this book.

- **Dr. Darshan Chaudhary**

TABLE OF CONTENTS

CHAPTER – 1

CHEMICAL CALCULATION

OBJECTIVES

After Studying this Chapter you will able to:

* *Know the method of finding formula weight of different compounds.*
* *Recognize the value of Avogadro number and its significance.*
* *Learn about the mole concept and the conversions of grams to moles.*
* *Know about the empirical and molecular formula and understand the method of arriving molecular formula from empirical formula.*
* *Understand the stoichiometric equation.*
* *Know about balancing the equation in its molecular form.*
* *Understand the concept of reduction and oxidation.*
* *Know about the method of balancing redox equation using oxidation number.*

1.1 Formula Weight (FW) or Formula Mass

The formula weight of a substance is the sum of the atomic weights of all atoms in a formula unit of the compound, whether molecular or not.

Sodium chloride, NaCl, has a formula weight of 58.44 amu (22.99 amu from Na plus 35.45 amu from Cl). NaCl is ionic, so strictly speaking the expression "molecular weight of NaCl" has no meaning. On the other hand, the molecular weight and the formula weight calculated from the molecular formula of a substance are identical.

Solved Problem

Calculate the formula weight of each of the following to three significant figures, using a table of atomic weight (AW): (a) chloroform $CHCl_3$ (b) Iron (III) sulphate $Fe_2(SO_4)_3$.

Solution

a. 1 x AW of C   =   12.0 amu

    1 x AW of H   =   1.0 amu

3 x AW of Cl = 3 x 35.45 =   <u>106.4 amu</u>

Formula weight of $CHCl_3$ =   <u>119.4 amu</u>

The answer rounded to three significant figures is 119 amu.

b. Iron(III)Sulfate

2 x Atomic weight of Fe   = 2 x 55.8   = 111.6 amu

3 x Atomic weight of S   = 3 x 32.1   = 96.3 amu

3 x 4 Atomic weight of O = 12 x 16 = 192.0 amu

Formula weight of $Fe_2(SO_4)_3$ = 399.9 amu

The answer rounded to three significant figures is 4.00×10^2 amu.

Problems for Practice

Calculate the formula weights of the following compounds

a. NO_2 b. glucose ($C_6H_{12}O_6$) c. NaOH d. $Mg(OH)_2$

e. methanol (CH_3OH) f. PCl_3 g. K_2CO_3

1.2 Avogadro's Number (N_A)

The number of atoms in a 12-g sample of carbon - 12 is called Avogadro's number (to which we give the symbol N_A).

Recent measurements of this number give the value 6.0221367×10^{23}, which is 6.023×10^{23}.

A mole of a substance contains Avogadro's number of molecules. A dozen eggs equals 12 eggs, a gross of pencils equals 144 pencils and a mole of ethanol equals 6.023×10^{23} ethanol molecules.

Significance

The molecular mass of SO_2 is 64 g mol^{-1}. 64 g of SO_2 contains 6.023×10^{23} molecules of SO_2. $2.24 \times 10^{-2} m^3$ of SO_2 at S.T.P. contains 6.023×10^{23} molecules of SO_2.

Similarly the molecular mass of CO_2 is 44 g mol^{-1}. 44g of CO_2 contains 6.023×10^{23} molecules of CO_2. $2.24 \times 10^{-2} m^3$ of CO_2 at S.T.P contains 6.023×10^{23} molecules of CO_2.

1.3 Mole concept

While carrying out reaction we are often interested in knowing the number of atoms and molecules. Sometimes, we have to take the atoms or molecules of different reactants in a definite ratio.

Eg. Consider the following reaction $2 H_2 + O_2 \rightarrow 2H_2O$

In this reaction one molecule of oxygen reacts with two molecules of hydrogen. So it would be desirable to take the molecules of H_2 and oxygen in the ratio 2:1, so that the reactants are completely consumed during the reaction. But atoms and molecules are so small in size that is not possible to count them individually.

In order to overcome these difficulties, the concept of mole was introduced. According to this concept number of particles of the substance is related to the mass of the substance.

Definition

The mole may be defined as the amount of the substance that contains as many specified elementary particles as the number of atoms in 12g of carbon - 12 isotope.

(i.e) one mole of an atom consists of Avogadro number of particles.

One mole	= 6.023×10^{23} particles
One mole of oxygen molecule	= 6.023×10^{23} oxygen molecules
One mole of oxygen atom	= 6.023×10^{23} oxygen atoms
One mole of ethanol	= 6.023×10^{23} ethanol molecules

In using the term mole for ionic substances, we mean the number of formula units of the substance. For example, a mole of sodium carbonate, Na_2CO_3 is a quantity containing 6.023×10^{23} Na_2CO_3 units. But eachformula unit of Na_2CO_3 contains $2 \times 6.023 \times 10^{23}$ Na^+ ions and one CO_3^{2-}ions and $1 \times 6.023 \times 10^{23}$ CO_3^{2-} ions.

When using the term mole, it is important to specify the formula of the unit to avoid any misunderstanding.

Eg. A mole of oxygen atom (with the formula O) contains 6.023×10^{23} Oxygen atoms. A mole of oxygen molecule (formula O_2) contains 6.023×10^{23} O_2 molecules (i.e) $2 \times 6.023 \times 10^{23}$ oxygen.

Molar mass

The molar mass of a substance is the mass of one mole of the substance. The mass and moles can be related by means of the formula.

$$\text{Molar mass} = \frac{\text{Mass}}{\text{mole}}$$

Eg. Carbon has a molar mass of exactly 12g/mol.

1.3.1 Mole Calculations

To find the mass of one mole of substance, there are two important things to know.

i.   How much does a given number of moles of a substance weigh?

ii. How many moles of a given formula unit does a given mass of substance contain.

Both of them can be known by using dimensional analysis.

To illustrate, consider the conversion of grams of ethanol, C_2H_5OH, to moles of ethanol. The molar mass of ethanol is 46.1 g/mol, So, we write

1 mol C_2H_5OH = 46.1 g of $C_2 H_5OH$

Thus, the factor converting grams of ethanol to moles of ethanol is 1mol C_2H_5OH/46.1g C_2H_5OH. To covert moles of ethanol to grams of ethanol, we simply convert the conversion factor (46.1 g C_2H_5OH/1 mol C_2H_5OH).

Again, suppose you are going to prepare acetic acid from 10.0g of ethanol, C_2H_5OH. How many moles of C_2H_5OH is this? you convert 10.0g C_2H_5OH to moles C_2H_5OH by multiplying by the appropriate conversion factor.

$$1 \text{ mol } C_2H_5OH$$

10.0g C_2H_5OHx_____

$$46.1 \text{ g } C_2H_5OH$$

$$= 0.217 \text{ mol } C_2H_5OH$$

1.3.2 Converting Moles of Substances to Grams
Solved Problems

1. ZnI_2, can be prepared by the direct combination of elements. A chemist determines from the amounts of elements that 0.0654 mol ZnI_2 can be formed.

Solution

The molar mass of ZnI_2 is 319 g/mol. (The formula weight is 319 amu, which is obtained by summing the atomic weight in the formula) Thus

$$319 \text{ g } ZnI_2$$

0.0654 mol ZnI_2 x _____

$$1 \text{ mol } ZnI_2$$

$$= 20.9 \text{ gm } ZnI_2$$

Problems for Practice

1. H_2O_2 is a colourless liquid. A concentrated solution of it is used as a source of oxygen for Rocket propellant fuels. Dilute aqueous solutions are used as a bleach. Analysis of a solution shows that it contains 0.909 mol H_2O_2 in 1.00 L of solution. What is the mass of H_2O_2 in this volume of solution?.

2. Boric acid, H_3BO_3 is a mild antiseptic and is often used as an eye wash. A sample contains 0.543 mol H_3BO_3. What is the mass of boric acid in the sample?.

3. CS_2 is a colourless, highly inflammable liquid used in the manufacture of rayon and cellophane. A sample contains 0.0205

mol CS_2. Calculate the mass of CS_2 in the sample.

Converting Grams of Substances to Moles

In the preparation of lead(II)chromate $PbCrO_4$, 45.6 g of lead(II)chromate is obtained as a precipitate. How many moles of $PbCrO_4$ is this?

The molar mass of $PbCrO_4$ is 323 g/mol (i.e) 1 mol $PbCrO_4$ = 323 g $PbCrO_4$

Therefore,

$$45.6 \text{ g } PbCrO_4 \times \frac{1 \text{ mol.}PbCrO_4}{323 \quad PbCrO_4}$$

= 0.141 mol $PbCrO_4$

Problems for Practice

1. Nitric acid, HNO_3 is a colourless, corrosive liquid used in the manufacture of Nitrogen fertilizers and explosives. In an experiment to develop new explosives for mining operations, a 28.5 g sample of HNO_3 was poured into a beaker. How many moles of HNO_3 are there in this sample of HNO_3?

Calculation of the Number of Molecules in a Given Mass Solved Problem

How many molecules are there in a 3.46 g sample of hydrogen chloride, HCl?

Note: The number of molecules in a sample is related to moles of compound (1 mol HCl = 6.023 x 10^{23} HCl molecules). Therefore if you first convert grams HCl to moles, then you can convert moles to number of molecules).

1.4 Calculation of Empirical Formula from Quantitative Analysis and Percentage composition

Empirical Formula

"An empirical formula (or) simplest formula for a compound is the formula of a substance written with the smallest integer subscripts".

For most ionic substances, the empirical formula is the formula of the compound. This is often not the case for molecular substances. For example, the formula of sodium peroxide, an ionic compound of Na^+ and O_2^{2-}, is Na_2O_2. Its empirical formula is NaO. Thus empirical formula tells you the ratio of numbers of atoms in the compound.

Steps for writing the Empirical formula

The percentage of the elements in the compound is determined by determined by the following steps.

i. Divide the percentage of each element by its atomic mass. This will give the relative number of moles of various elements present in the compound.

ii. Divide the quotients obtained in the above step by the smallest of them so as to get a simple ratio of moles of various elements.

iii. Multiply the figures, so obtained by a suitable integer of necessary in order to obtain whole number ratio.

iv. Finally write down the symbols of the various elements side by side and put the above numbers as the subscripts to the lower right hand of each symbol. This will represent the empirical formula of the compound.

Solved Problem

A compound has the following composition $Mg = 9.76\%, S = 13.01\%$, $O = 26.01, H_2O = 51.22$, what is its empirical formula?

[Mg = 24, S = 32, O = 16, H = 1]

Solution

Element	%	Relative No. of Moles	Simple ratio moles	Simplest whole No. ratio
Magnesium	9.76	$\dfrac{9.76}{24} = 0.406$	$\dfrac{0.406}{0.406} = 1$	1
Sulphur	13.01	$\dfrac{13.01}{32} = 0.406$	$\dfrac{0.406}{0.406} = 1$	1
Oxygen	26.01	$\dfrac{26.01}{16} = 1.625$	$\dfrac{1.625}{0.406} = 4$	4
Water	51.22	$\dfrac{51.22}{18} = 2.846$	$\dfrac{2.846}{0.406} = 7$	7

Hence the empirical formula is $MgSO_4.7H_2O$.

Problems for Practice

1. A substance on analysis, gave the following percentage composition, Na = 43.4%, C = 11.3%, 0 = 43.3% calculate its empirical formula [Na = 23, C = 12, O = 16].
Ans:- Na_2CO_3

2. What is the simplest formula of the compound which has the following percentage composition: Carbon 80%, hydrogen 20%.
Ans:- CH_3

3. A compound on analysis gave the following percentage composition: C - 54.54%, H = 9.09%, 0 = 36.36%
Ans:- C_2H_4O

1.4.1 Molecular Formula from Empirical Formula

The molecular formula of a compound is a multiple of its empirical formula.

Example

The molecular formula of acetylene, C_2H_2 is equivalent to $(CH)_2$, and the molecular formula of benzene, C_6H_6 is equivalent to $(CH)_6$. Therefore, the molecular weight is some multiple of the empirical formula weight, which is obtained by summing the atomic Weights from the empirical formula. For any molecular compound.

Molecular Weight = n x empirical formula weight.

Where `n' is the whole number of empirical formula units in the molecule. The molecular formula can be obtained by multiplying the subscripts of the empirical formula by `n' which can be calculated by the following equation

$$n = \frac{\text{Molecular Weight}}{\text{Empirical formula Weight}}$$

Steps for writing the molecular formula

i. Calculate the empirical formula

ii. Find out the empirical formula mass by adding the atomic mass of all the atoms present in the empirical formula of the compound.

iii. Divide the molecular mass (determined experimentally by some suitable method) by the empirical formula mass and find out the value of n which is a whole number.

iv. Multiply the empirical formula of the compound with n, so as to find out the molecular formula of the compound.

Solved Problem

1. A compound on analysis gave the following percentage composition C = 54.54%, H, 9.09% 0 = 36.36. The vapour density of the compound was found to be 44. Find out the molecular formula of the compound.

Solution

Calculation of empirical formula

Element	%	Relative No. of Moles	Simple ratio Moles	Simplest whole No. ratio
C	54.54	$\dfrac{54.54}{12} = 4.53$	$\dfrac{4.53}{2.27} = 2$	2
H	9.09	$\dfrac{9.09}{1} = 9.09$	$\dfrac{9.09}{2.27} = 4$	4
O	36.36	$\dfrac{36.36}{16} = 2.27$	$\dfrac{2.27}{2.27} = 1$	1

Empirical formula is $C_2 H_4 O$.

Calculation of Molecular formula

Empirical formula mass $= 12 \times 2 + 1 \times 4 + 16 \times 1 = 44$

Molecular mass $= 2 \times$ Vapour density

$= 2 \times 44 = 88$

$$n = \frac{\text{Molecular mass}}{\text{Empirical Formula mass}} = \frac{88}{44} = 2$$

Molecular formula = Empirical formula x n

$= C_2 H_4 O \times 2$

$$= C_4 H_8 O_2$$

2. A compound on analysis gave the following percentage composition: Na=14.31% S = 9.97%, H = 6.22%, O = 69.5%, calculate the molecular formula of the compound on the assumption that all the hydrogen in the compound is present in combination with oxygen as water of crystallization. Molecular mass of the compound is 322 [Na = 23, S = 32, H = 1, 0 = 16].

Solution:- Calculation of empirical formula

Element	%	Relative No. of moles	Simple ratio Moles	Simplest whole No. ratio
Na	14.31	$\dfrac{14.31}{23} = 0.62$	$\dfrac{0.62}{0.31} = 2$	2
S	9.97	$\dfrac{19.97}{32} = 0.31$	$\dfrac{0.31}{0.31} = 1$	1
H	6.22	$\dfrac{6.22}{1} = 6.22$	$\dfrac{6.22}{0.31} = 20$	20
O	69.5	$\dfrac{69.5}{16} = 4.34$	$\dfrac{4.34}{0.31} = 14$	14

The empirical formula is $Na_2 SH_{20} O_{14}$

Since all hydrogens are present as H_2O in the compound, it means 20 hydrogen atoms must have combined. It means 20 hydrogen atoms must have combined with 10 atoms of oxygen to form 10 molecules of water of crystallization. The remaining (14 - 10 = 4) atoms of oxygen should be present with the rest of the compound.

Hence, molecular formula = $Na_2SO_4.10H_2O$.

Problems for Practice

1. An organic compound was found to have contained carbon = 40.65%, hydrogen = 8.55% and Nitrogen = 23.7%. Its vapour - density was found to be 29.5. What is the molecular formula of the compound?

Ans:- C_2H_5NO

2. A compound contains 32% carbon, 4% hydrogen and rest oxygen. Its vapour density is 75. Calculate the empirical and molecular formula.

Ans:- $C_2H_3O_3$, $C_4H_6O_6$

3. An acid of molecular mass 104 contains 34.6% carbon, 3.85% hydrogen and the rest is oxygen. Calculate the molecular formula of the acid.

4. What is the simplest formula of the compound which has the following percentage composition: carbon 80%, Hydrogen 20%, If the molecular mass is 30, calculate its molecular formula.

1.5 Stoichiometry Equations

Stoichiometry

Stoichiometry is the calculation of the quantities of reactants and products involved in the chemical reaction. It is the study of the relationship between the number of mole of the reactants and products of a chemical reaction. A stoichiometric equation is a short scientific representation of a chemical reaction.

Rules for writing stoichiometric equations

i. In order to write the stoichiometric equation correctly, we must know the reacting substances, all the products formed and their chemical formula.

ii. The formulae of the reactant must be written on the left side of arrow with a positive sign between them.

iii. The formulae of the products formed are written on the right side of the arrow mark. If there is more than one product, a positive sign is placed between them. The equation thus obtained is called skeleton equation. For example, the Chemical reaction between Barium chloride and sodium sulphate producing $BaSO_4$ and $NaCl$ is represented by the equation as

$$BaCl_2 + Na_2SO_4 \rightarrow BaSO_4 + NaCl$$

This skeleton equation itself is a balanced one. But in many cases the skeleton equation is not a balanced one.

For example, the decomposition of Lead Nitrate giving Lead oxide, NO_2 and oxygen. The skeletal equation for this reaction is

$$Pb(NO_3)_2 \rightarrow PbO + NO_2 + O_2$$

iv. In the skeleton equation, the numbers and kinds of particles present on both sides of the arrow are not equal.

v. During balancing the equation, the formulae of substances should not be altered, but the number of molecules with it only be suitably changed.

vi. Important conditions such as temperature, pressure, catalyst etc., may be noted above (or) below the arrow of the equation.

vii. An upward arrow ($\rightarrow$) is placed on the right side of the formula of a gaseous product and a downward arrow ($\rightarrow$) on the right side of the formulae of a precipitated product.

viii. All the reactants and products should be written as molecules including the elements like hydrogen, oxygen, nitrogen, fluorine chlorine, bromine and iodine as H_2, O_2, N_2, F_2, Cl_2, Br_2 and I_2.

1.5.1 Balancing chemical equation in its molecular form

A chemical equation is called balanced equation only when the numbers and kinds of molecules present on both sides are equal. The several steps involved in balancing chemical equation are discussed below.

Example 1

Hydrogen combines with bromine giving HBr

$$H_2 + Br_2 \rightarrow HBr$$

This is the skeletal equation. The number of atoms of hydrogen on the left side is two but on the right side it is one. So the number of molecules of HBr is to be multiplied by two. Then the equation becomes

$$H_2 + Br_2 \rightarrow 2HBr$$

This is the balanced (or) stoichiometric equation.

Example 2

Potassium permanganate reacts with HCl to give KCl and other products. The skeletal equation is

$$KMnO_4 + HCl \rightarrow KCl + MnCl_2 + H_2O + Cl_2$$

If an element is present only one substance in the left hand side of the equation and if the same element is present only one of the substances in the right side, it may be taken up first while balancing the equation.

According to the above rule, the balancing of the equation may be started with respect to K, Mn, O (or) H but not with Cl.

There are

L.H.S. R.H.S

K = 1   1

Mn = 1 1

O = 4   1

So the equation becomes

$$KMnO_4 + HCl \rightarrow KCl + MnCl_2 + 4H_2O + Cl_2$$

Now there are eight hydrogen atoms on the right side of the equation, we must write 8 HCl.

$$KMnO_4 + 8HCl \rightarrow KCl + MnCl_2 + 4H_2O + Cl_2$$

Of the eight chlorine atoms on the left, one is disposed of in KCl and two in $MnCl_2$ leaving five free chlorine atoms. Therefore, the above equation becomes

$$KMnO_4 + 8HCl \rightarrow KCl + MnCl_2 + 4H_2O + 5/2\ Cl_2$$

Equations are written with whole number coefficient and so the equation is multiplied through out by 2 to become

$$2KMnO_4 + 16HCl \rightarrow 2KCl + 2MnCl_2 + 8H_2O + 5Cl_2$$

1.5.2 Redox reactions [Reduction - oxidation]

In our daily life we come across process like fading of the colour of the clothes, burning of the combustible substances such as cooking gas, wood, coal, rusting of iron articles, etc. All such processes fall in the category of specific type of chemical reactions called reduction - oxidation (or) redox reactions. A large number of industrial processes like, electroplating, extraction of metals like aluminum and sodium, manufactures of caustic soda, etc., are also based upon the redox reactions. Redox reactions also form the basis of electrochemical and electrolytic cells. According to the classical concept, oxidation and reduction may be defined as,

Oxidation is a process of addition of oxygen (or) removal of hydrogen

Reduction is a process of removal of oxygen (or) addition of hydrogen.

Example

Reaction of Cl_2 and H_2S

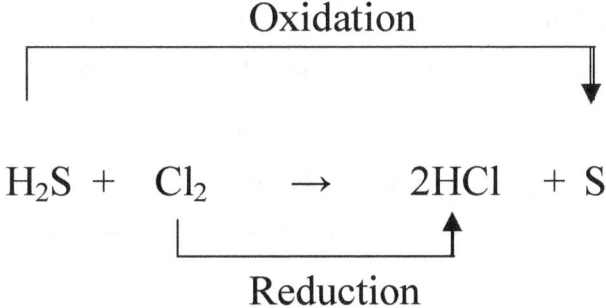

In the above reaction, hydrogen is being removed from hydrogen sulphide (H_2S) and is being added to chlorine (Cl_2). Thus, H_2S is oxidized and Cl_2 is reduced.

Electronic concept of oxidation and Reduction

According to electronic concept, oxidation is a process in which an atom taking part in chemical reaction loses one or more electrons. The loss of electrons results in the increase of positive charge (or) decrease of negative of the species. For example.

$$Fe^{2+} \rightarrow Fe^{3+} + e^- \text{ [Increase of positive charge]}$$
$$Cu \rightarrow Cu^{2+} + 2e^- \text{ [Increase of positive charge]}$$

The species which undergo the loss of electrons during the reactions are called reducing agents or reductants. Fe^{2+} and Cu are reducing agents in the above example.

Reduction

Reduction is a process in which an atom (or) a group of atoms taking part in chemical reaction gains one (or) more electrons. The gain of electrons result in the decrease of positive charge (or) increase of negative charge of the species. For example,

$$Fe^{3+} + e^- \rightarrow Fe^{2+} \text{ [Decrease of positive charges]}$$
$$Zn^{2+} + 2e^- \rightarrow Zn \text{ [Decrease of positive charges]}$$

The species which undergo gain of electrons during the reactions are called oxidizing agents (or) oxidants. In the above reaction, Fe^{3+} and Zn^{2+} are oxidizing agents.

Oxidation Number (or) Oxidation State

Oxidation number of the element is defined as the residual charge which its atom has (or) appears to have when all other atoms from the molecule are removed as ions.

Atoms can have positive, zero or negative values of oxidation numbers depending upon their state of combination.

General Rules for assigning Oxidation Number to an atom

The following rules are employed for determining oxidation number of the atoms.

1. The oxidation number of the element in the free (or) elementary state is always Zero.

    Oxidation number of Helium in He = 0

    Oxidation number of chlorine in $Cl_2 = 0$

2. The oxidation number of the element in monoatomic ion is equal to the charge on the ion.

3. The oxidation number of fluorine is always - 1 in all its compounds.

4. Hydrogen is assigned oxidation number +1 in all its compounds except in metal hydrides. In metal hydrides like NaH, MgH_2, CaH_2, LiH, etc., the oxidation number of hydrogen is -1.

5. Oxygen is assigned oxidation number -2 in most of its compounds, however in peroxides like H_2O_2, BaO_2, Na_2O_2, etc its oxidation number is -

Similarly the exception also occurs in compounds of Fluorine and oxygen like OF_2 and O_2F_2 in which the oxidation number of oxygen is +2 and +1 respectively.

6. The oxidation numbers of all the atoms in neutral molecule is Zero. In case of polyatomic ion the sum of oxidation numbers of all its atoms is equal to the charge on the ion.
7. In binary compounds of metal and non-metal the metal atom has positive oxidation number while the non-metal atom has negative oxidation number. Example. Oxidation number of K in KI is +1 but oxidation number of I is - I.
8. In binary compounds of non-metals, the more electronegative atom has negative oxidation number, but less electronegative atom has positive oxidation number. Example: Oxidation number of Cl in ClF_3 is positive (+3) while that in IC_l is negative (-1).

Problem

Calculate the oxidation number of underlined elements in the following species.

$\underline{C}O_2$, $\underline{Cr}_2O_7^{2-}$, $\underline{Pb}_3O_4$, $\underline{P}O_4^{3-}$

Solution

1. C in CO_2. Let oxidation number of C be x. Oxidation number of each O atom = -2. Sum of oxidation number of all atoms = x+2 (-2) x-4.

   As it is neutral molecule, the sum must be equal to zero.

   x-4 = 0 (or) x = + 4

2. Cr in $Cr_2O_7^{2-}$ Let oxidation number of Cr = x. Oxidation number of each oxygen atom =-2. Sum of oxidation number of all atoms

   $$2x + 7(-2) = 2x - 14$$

   Sum of oxidation number must be equal to the charge on the ion.

   Thus, 2x - 14 = -2

   $$2x = +12$$
   $$x = 12/2$$
   $$x = 6$$

Problems for Practice

Calculate the oxidation number of underlined elements in the following species.

a. $\underline{Mn}SO4$ b. $\underline{S}_2O_3$ c. $H\underline{N}O_3$ d. $K_2\underline{Mn}O_4$ e. $\underline{N}H_4^+$

Basics of Inorganic Chemistry

Oxidation and Reduction in Terms of Oxidation Number

Oxidation

"A chemical process in which oxidation number of the element increases".

Reduction

"A chemical process in which oxidation number of the element decreases".

Eg.Reaction between H_2S and Br_2 to give HBr and Sulphur.

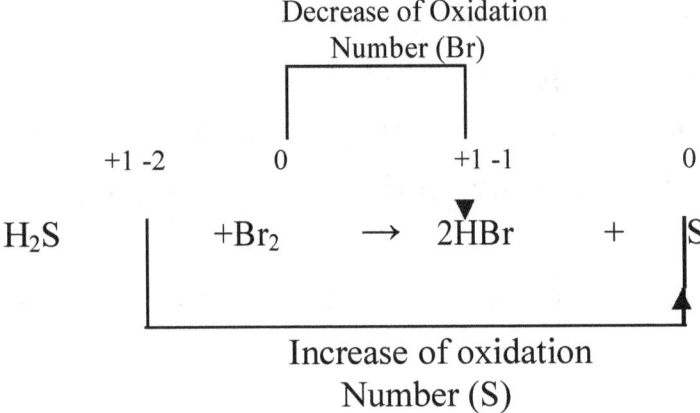

In the above reaction, the oxidation number of bromine decreases from 0 to -1, thus it is reduced. The oxidation number of S increases from -2 to 0. Hence H_2S is oxidized.

Under the concept of oxidation number, oxidizing and reducing agent can be defined as follows.

i. Oxidizing agent is a substance which undergoes decrease in the oxidation number of one of its elements.

ii. Reducing agent is a substance which undergoes increase in the oxidation number of one of its elements.

In the above reaction H_2S is reducing agent while Br_2 is oxidizing agent.

Solved Problem

Identify the oxidizing agent, reducing agent, substance oxidized and substance reduced in the following reactions.

$$MnO_2 + 4HCl \rightarrow MnCl_2 + Cl_2 + 2H_2O$$

Solution

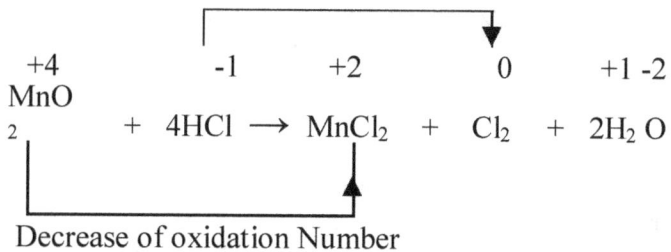

As it is clear, manganese decrease its oxidation number from +4 to +2. Hence, MnO_2 gets reduced and it is an oxidizing agent. Chlorine atom in HCl increases its oxidation number from -1 to 0. Thus, HCl gets oxidized and it is reducing agent.

Balancing Redox reaction by Oxidation Number Method

The various steps involved in the balancing of redox equations according to this method are:

1. Write skeleton equation and indicate oxidation number of each element and thus identify the elements undergoing change in oxidation number.

2. Determine the increase and decrease of oxidation number per atom. Multiply the increase (or) decrease of oxidation number of atoms undergoing the change.

3. Equalize the increase in oxidation number and decrease in oxidation number on the reactant side by multiplying the respective formulae with suitable integers.

4. Balance the equation with respect to all atoms other than O and H atoms.

5. Balance oxygen by adding equal number of water molecules to the side falling short of oxygen atoms.

6. H atoms are balanced depending upon the medium in same way as followed in ion electron method.

Let us balance the following equations by oxidation number method.

$$MnO_2 + \underline{Cl^-} \rightarrow Mn^{2+} + Cl_2 + H_2O \text{ in acidic}$$
$$\text{medium}$$

Step 1

$$MnO_2 + Cl^- \rightarrow Mn^{2+} + Cl_2 + H_2O$$

Step 2

O.N. Decreases by 2 per Mn

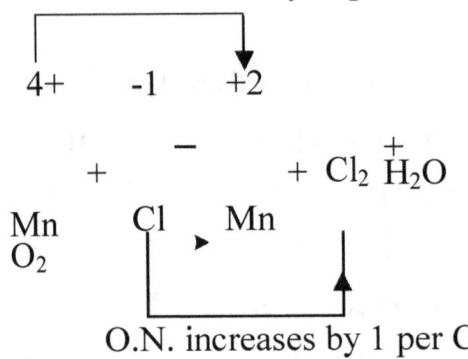

O.N. increases by 1 per Cl

Step 3

Equalize the increase/decrease in O.N by multiply MnO_2 by 1 and Cl^{-1} by 2.

$$MnO_2 + 2Cl^- \rightarrow Mn^{2+} + Cl_2 + H_2O$$

Step 4

Balance other atoms except H and O. Here they are all balanced.

Step 5

Balance O atoms by adding H_2O molecules to the side falling short of oxygen atoms.

$$MnO_2 + 2Cl^- \rightarrow Mn^{2+} + Cl_2 + H_2O + H_2O$$

Step 6

Balance H atoms by adding H^+ ions to the side falling short of H atoms

$$MnO_2 + 2Cl^- + 4H^+ \rightarrow Mn^{2+} + Cl_2 + 2H_2O$$

Problems for practice

Balance the following equations

1. $Mg + NO_3^- \rightarrow Mg^{2+} + N_2O + H_2O$
   (in acidic medium)

2. $Cr^{3+} + Na_2O_2 \rightarrow CrO_4^- + Na^+$

3. $S^{2-} + NO_3^- \rightarrow NO + S$

4. $FeS + O_2 \rightarrow Fe_2O_3 + SO_2$ (molecular form)

5. $Cl_2 + OH^- \rightarrow Cl^- + ClO_3^- + H_2O$

1.6 Calculations based on chemical equations

Stoichiometric problems are solved readily with reference to the equation describing the chemical change. From a stoichiometric chemical equation, we know how many molecules of reactant react and how many molecules of products are formed. When the molecular mass of the substances are inserted, the equation indicates how many parts by mass of reactants react to produce how many parts by mass of products. The parts by mass are usually in kg. So it is possible to calculate desired mass of the product for a known mass of the reactant or vice versa.

1.6.1 Mass / Mass Relationship

Example 1

Calculate the mass of CO_2 that would be obtained by completely dissolving 10kg of pure $CaCO_3$ in HCl.

$$CaCO_3 + 2HCl \rightarrow CaCl_2 + H_2O + CO_2$$
$$100 \text{ kg x } 10^{-3} \qquad\qquad 44 \text{ kg x } 10^{-3}$$

100 kg of $CaCO_3$ produces 44×10^{-3} kg of CO_2

10 kg of $CaCO_3$ produces $= \dfrac{44 \times 10^{-3}}{100 \times 10^{-3}} \times 10$

$= 4.4$ kg of CO_2

Example 2

Calculate the mass of oxygen obtained by complete decomposition of 10kg of pure potassium chlorate (Atomic mass K=39, O=16 and Cl = 35.5)

$$2KClO_3 \rightarrow 2 KCl + 3O_2$$

Molecular mass of KClO3 = 39+35.5+48=122.5

Molecular Mass of O_2= 16 + 16 = 32

According to the Stoichiometric equation written above (2 x 122.5) x 10^{-3} kg of $KClO_3$ on heating gives (3 x 32) x 10^{-3} kg of oxygen.

$$\text{of } KClO_3 \text{ gives} = \frac{3 \times 32 \times 10^{-3} \; 10kg}{2 \times 122.5 \times 10^{-3}} \times 10$$

$$= 3.92 \text{ kg of } O_2$$

Example 3

Calculate the mass of lime that can be prepared by heating 200 kg of limestone that is 90% pure $CaCO_3$

$$CaCO_3 \longrightarrow CaO + CO_2$$
$$100 \text{ kg} \times 10^{-3} \quad 56 \text{ kg} \times 10^{-3}$$

$$200 \text{ kg of } 90\% \text{ pure } CaCO_3 = 200 \times \frac{90}{100}$$
$$= 180 \text{ kg pure } CaCO_3$$

100×10^{-3} kg of pure $CaCO_3$ on heating gives 56×10^{-3} kg of CaO

$$\begin{array}{l}
180 \text{ kg of } CaCO_3 \\
\text{gives on heating}
\end{array} = \frac{56 \times 10^{-3} \times 180}{100 \times 10^{-3}}$$

$$= 100.8 \text{ kg CaO}$$

Methods of Expressing the concentration of solution

The concentration of a solution refers to the amount of solute present in the given quantity of solution or solvent. The concentration of a solution may be expressed quantitatively in any of the following ways.

1. Strength

The Strength of a solution is defined as the amount of the solute in grams, present in one litre of the solution. It is expressed in g L^{-1}.

$$\text{Strength} = \frac{\text{Mass of solute in grams}}{\text{Volume of solution in litres}}$$

If X gram of solute is present in V cm^3 of a given solution then

$$\text{Strength} = \frac{X \times 1000}{V}$$

2. Molarity (M)

Molarity of a solution is defined as the number of gram-moles of solute dissolved in 1 litre of a solution

$$\text{Molarity} = \frac{\text{No. of moles of solute}}{\text{Volume of Solution in litres}}$$

If `X' grams of the solute is present in V cm^3 of a given solution, then,

$$\text{Molarity} = \frac{X}{\text{Mol. mass}} \times \frac{1000}{V}$$

Molarity is represented by the symbol M. Molarity can also be calculated from the strength as follows

$$\text{Molarity} = \frac{\text{Strength in grams per litre}}{\text{Molecular mass of the solute}}$$

Example

A 0.1M solution of Sugar, $C_{12}H_{22}O_{11}$ (mol.mass = 342), means that 34.2 g of sugar is present in one litre (1000 cm^3) of the solution.

3. Normality

Normality of a solution is defined as the number of gram equivalents of the solute dissolved per litre of the given solution.

$$\text{Normality} = \frac{\text{Number of gram-equivalents of solute}}{\text{Volume of Solution in litre}}$$

If X grams of the solute is present in V cm^3 of a given solution, then,

$$\text{Normality} = \frac{X}{\text{Eq.mass}} \times \frac{1000 \text{ mL}}{V}$$

Normality is represented by the symbol N. Normality can also be calculated from strength as follows

$$\text{Normality} = \frac{\text{Strength in grams per litre}}{\text{Eq.mass of the solute}}$$

Example

A 0.1N (or decinormal) solution of H_2SO_4 (Eq.mass = 49), means that 4.9 g of H_2SO_4 is present in one litre (1000 cm^3) of the solution.

4. Molality (m)

Molality of a solution is defined as the number of gram-moles of solute dissolved in 1000 grams (or 1 kg) of a Solvent. Mathematically,

$$\text{Molality} = \frac{\text{Number of moles of solute}}{\text{Mass of solvent in kilograms}}$$

"If X grams of the solute is dissolved in b grams of the solvent", then

$$\text{Molality} = \frac{X}{\text{Mol. mass}} \times \frac{1000g}{bg}$$

Molality is represented by the symbol 'm'.

Example

A 0.1m Solution of glucose $C_6H_{12}O_6$ (Mol.mass = 180), means that 18g of glucose is present in 1000g (or one kilogram) of water.

5. Mole Fraction

Mole fraction is the ratio of number of moles of one component (Solute or Solvent) to the total number of moles of all the components (Solute and Solvent) present in the Solution. It is denoted by X. Let us suppose that a solution contains 2 components A&B and suppose that n_A moles of A and n_B moles of B are present in the solution. Then,

$$\text{Mole fraction of A, } X_A = \frac{n_A}{n_A + n_B} \quad\text{.....(1)}$$

$$\text{Mole fraction of B, } X_B = \frac{n_B}{n_A + n_B} \quad\text{......(2)}$$

Adding 1 & 2 we get,

$$X_A + X_B = \frac{n_A}{n_A + n_B} + \frac{n_B}{n_A + n_B} = \frac{n_A + n_B}{n_A + n_B}$$

Thus, sum of the two mole fractions is one. Therefore, if mole fraction of one component in a binary solution is known, that of the other can be calculated.

1.8 Calculations based on Principle of Volumetric Analysis
1.8.1 Volumetric Analysis

An important method for determining the amount of a particular substance is based on measuring the volume of reactant solution. Suppose substance A in solution reacts with substance B. If you know the volume and concentration of a solution of B that just reacts with substance A in a sample, you can determine the amount of A.

Titration is a procedure for determining the amount of substance A by adding a carefully measured volume of a solution of A with known concentration of B until the reaction of A and B is just completed. Volumetric analysis is a method of analysis based on titrations.

Law

"Equal volumes of equinormal solutions exactly neutralize the other solution having same concentration and volume".

$$V_1 N_1 = V_2 N_2$$

V_1, V_2 - Volume of solutions.

N_1, N_2 - Strength of solutions.

1.8.2 Determination of equivalent masses of elements

Equivalent masses can be determined by the following methods:
1. Hydrogen displacement method
2. Oxide method
3. Chloride method
4. Metal displacement method

Hydrogen displacement method

This method is used to determine the equivalent mass of those metals such as magnesium, zinc and aluminum which react with dilute acids and readily displace hydrogen.

$Mg + 2HCl \rightarrow MgCl_2 + H_2$

$Zn + H_2SO_4 \rightarrow ZnSO_4 + H_2$

$2Al + 6HCl \rightarrow 2AlCl_3 + 3H_2$

From the mass of the metal and the volume of hydrogen displaced, the equivalent mass of the metal can be calculated.

Problem 1

0.548 g of the metal reacts with dilute acid and liberates 0.0198 g of hydrogen at S.T.P. Calculate the equivalent mass of the metal.

0.548 g of the metal displaces 0.0198 g of hydrogen

The mass of the metal which will displace

1.008 g of hydrogen $= \dfrac{1.008 \times 0.548}{0.0198}$ g of metal

The equivalent mass of the metal $= 27.90$ g equiv^{-1}

Oxide Method

This method is employed to determine the equivalent mass of those elements which readily form their oxides eg. magnesium, copper etc. Oxide of an element can be formed by direct method or by indirect method.

Magnesium forms its oxide directly on heating

$$2Mg + O_2 \rightarrow 2MgO$$

In the case of copper, its oxide is obtained in an indirect manner i.e. copper is dissolved in concentrated nitric acid and the copper(II) nitrate formed after evaporation is strongly heated to give copper (II) oxide.

$Cu + 4HNO_3 \rightarrow Cu(NO_3)_2 + 2H_2O + 2NO_2$

$2Cu(NO_3)_2 \rightarrow 2CuO + 4NO_2 + O_2$

Calculations

Mass of the element taken $= w_1$ g

Mass of the oxide of the element $= w_2$ g

Mass of oxygen $= (w_2 - w_1)$ g

$(w_2 - w_1)$ g of oxygen has combined with w_1 g of the metal.

$\therefore$ 8 g of oxygen will combine with $\dfrac{w_1}{w_2 - w_1} \times 8$

This value represents the equivalent mass of the metal.

Chloride Method

The equivalent mass of those elements which readily form their chlorides can be determined by chloride method. For example, a known mass of pure silver is dissolved completely in dilute nitric acid. The resulting silver nitrate solution is treated with pure hydrochloric acid when silver chloride is precipitated. It is then filtered, dried and weighed. Thus from the masses of the metal and its chloride, the equivalent mass of the metal can be determined as follows:

Calculations

Mass of the metal $= w_1$ g

Mass of the metal chloride $= w_2$ g

Mass of chlorine $= (w_2 - w_2)$ g

$(w_2 - w_1)$ g of chlorine has combined with w_1 of the metal

35.46 g of chlorine will combine with

$$\frac{35.46 \times w_1}{(w_2 - w_1)} \text{ g of the metal}$$

This value gives the equivalent mass of the metal.

Uses of volumetric laws

If the volume of the acid is represented as V_1, the normality of the acid as N_1, volume of base as V_2 the normality of the base as N_2, then from the law of volumetric analysis it follows that

$$V_1 \times N_1 = V_2 \times N_2$$

All volumetric estimations are based on the above relationship. If any three quantities are known, the fourth one can readily be calculated using the above expression.

1.8.3 Equivalent mass of acid, base, salt, oxidizing agent and reducing agent

Acids contain one or more replaceble hydrogen atoms. The number of replaceable hydrogen atoms present in a molecule of the acid is referred to its basicity.

Equivalent mass of an acid is the number of parts by mass of the acid which contains 1.008 part by mass of replaceable hydrogen atom.

$$\text{Equivalent} = \frac{\text{molar mass of the acid}}{\text{No. of replaceble hydrogen atom}}$$

mass of an acid

(or)

$$\text{molar mass of the acid} = \frac{}{\text{basicity of the acid}}$$

For example, the basicity of sulphuric acid is 2.

1. Equivalent mass of the base

Equivalent mass of a base is the number of parts by mass of the basewhich contains one replaceable hydroxyl ion or which completely neutralizes one gram equivalent of an acid. The number of hydroxyl ions present in one mole of a base is known as the acidity of the base. Sodium hydroxide, potassium hydroxide, ammonium hydroxide are examples of monoacidic bases.

Calcium hydroxide is a diacritic base

In general,

$$\text{equivalent mass of a base} = \frac{\text{molar mass of the base}}{\text{acidity of the base}}$$

Equivalent mass of KOH = 56 /1 = 56

3. Equivalent mass of a salt

Equivalent mass of a salt is a number of parts by mass of the salt that is produced by the neutralization of one equivalent of an acid by a base.

In the case of salt like potassium chloride, the salt formed by the neutralization of one equivalent of an acid by a base.

$$KOH + HCl \rightarrow KCl + H_2O$$

Therefore, the equivalent mass of the salt is equal to its molar mass.

4. Equivalent mass of an oxidizing agent

The equivalent mass of an oxidizing agent is the number of parts by mass which can furnish 8 parts by mass of oxygen for oxidation either directly or indirectly.

For example, potassium permanganate is an oxidizing agent. In acid medium potassium permanganate reacts as follows

$$2KMnO_4 + 3H_2SO_4 \rightarrow K_2SO_4 + 2MnSO_4 + 3H_2O + 5[O]\ 316$$
$$80$$

80 parts by mass of oxygen are made available from 316 parts by mass of $KMnO_4$

8 parts by mass of oxygen will be furnished by

$$\frac{316 \times 8}{80} = 3.16$$

Equivalent mass of $KMnO_4$= 31.6g equiv^{-1}

5. Equivalent mass of a reducing agent

The equivalent mass of a reducing agent is the number of parts by mass of the reducing agent which is completely oxidized by 8 parts by mass of oxygen or with one equivalent of any oxidizing agent.

(i) Ferrous sulphate reacts with an oxidizing agent in acid medium according to the equation

$$2\ FeSO_4 + H_2SO_4 + (O) \rightarrow Fe_2(SO_4)_3 + H_2O$$
$$2 \times 152g \qquad 16g$$

16 parts by mass of oxygen oxidized 304 parts by mass of ferrous sulphate

8 parts by mass of oxygen will oxidize 304/16 x 8 parts by mass of ferrous sulphate.

(ii) In acid medium, oxalic acid is oxidized according to the equation

$$(COOH)_2 + (O) \rightarrow 2CO_2 + H_2O$$

16 Parts by mass of oxygen oxidized 90 parts by mass of anhydrous oxalic acid.

8 parts by mass of oxygen will oxidize 90/16 x 8 = 45 parts by mass of anhydrous oxalic acid.

Equivalent mass of anhydrous oxalic acid = 45 g equiv^{-1}

But equivalent mass of crystalline oxalic acid, $(COOH)_2$.
$2H_2O$=126/2 = 63 g equiv^{-1}.

1.8.4 Determination of Molecular Mass Victor-Meyer's Method Principle

In this method a known mass of a volatile liquid or solid is converted into its vapour by heating in a Victor-Meyer's tube. The vapour displaces its own volume of air. The volume of air displaced by the vapour is measured at the experimental temperature and pressure. The volume of the vapour at s.t.p is then calculated. From this the mass of $2.24 \times 10^{-2} m^3$ of the vapour at S.T.P. is calculated. This value represents the molecular mass of the substance.

SUMMARY

SI units and different scientific notation. Molecular mass, Mole concept, Avogadro number and its significance are dealt. The application of the various concepts are explained by solving problems. By knowing the percentage composition of elements in a compound, empirical formula and molecular formula can be calculated.

It is important to write the stoichiometric equation. So, the method of balancing the any equation explained and given or practice. And also the method of balancing redox equation using oxidation number is dealt.

REFERENCES:

1. General Chemistry – John Russell
   McGraw Hill International Editions 3rd Edition.
2. University General Chemistry
   An Introduction to Chemical Science edited by CNR Rao. McMillan Indian Limited, Reprint-2002.
3. Heinemann Advanced Science Chemistry – Second Edition Ann and Patrick Fullick 2000 Heineman Educational Publishers, Oxford.
4. Inorganic Chemistry, P.L. Soni.

CHAPTER – 2

GENERAL INTRODUCTION TOMETALLURGY
OBJECTIVES
After Studying this Chapter you will able to:
· *Know the ores and minerals of elements*
· *Learn the purification methods of ores*
· *Understand the different metallurgical processes*
· *Know the importance of purification of metals*
· *Understand clearly the extraction of Cu, Au, Ag, Pb, Zn and Al*
· *Gain knowledge about the mineral wealth of India and Tamilnadu*

2.1 Ores and minerals

Metals occur in nature sometimes in **free or native state,** but most of these occur in nature in the form of **chemical combination,** i.e., in the form of their stable compounds which are associated with **gangue ormatrix.** The earthy impurities such as sand, clay, rocks etc. associatedwith ore are collectively known as **gangue or matrix.** Thus a large number of metals in nature occur in the combined form along with other elements, but some metals, such as Ag, Au, Pt etc. occur in the **nativeform (metallic state)** in some regions. Ag occurs in **native (or free)** aswell as in the form of **compounds.** The natural material in which the metal or their compounds occur in the earth is known as **mineral.** Thus a mineral is a naturally occurring material present in earth's crust which contains metal in the native (or free state) or in combined state.

A mineral may be a single compound or a complex mixture of various compounds. When a mineral contains sufficient amount of a metal in combined state, from which it can be readily and profitably separated on commercial scale, then the mineral is said to be an ore of the metal. **Thus all ores are minerals, but all minerals are not ores.** A mineral from which a metal can be profitably extracted is called an **ore.** For example, clay ($Al_2O_3 2SiO_2.2H_2O$) and bauxite ($Al_2O_3.2H_2O$) are two minerals of aluminium, but aluminium, can be profitably extracted only from bauxite and not from the clay. **Hence bauxite is an ore ofaluminium, while clay is a mineral.** The biggest source of metal is the earth's crust and the process of taking out the ores from the earth crust is called **mining.**

In the combined state ores are generally found in the form of oxides, sulphides, carbonates, sulphates, chlorides and silicates. The important ores are given in Table 2.1.

Table 2.1 Classification of ores

Ore	Ore or Mineral	Composition	Metal Present
Oxide ores	Bauxite	$Al_2O_3 .2H_2O$	Al
	Cuprite	Cu_2O	Cu
	Haematite	Fe_2O_3	Fe
	Zincite	ZnO	Zn
	Tinstone or Casseterite	SnO_2	Sn
	Pyrolusite	MnO_2	Mn
	Pitch Blende	U_3O_8	U
Sulphide Ores	Copper Pyrites	Cu_2S, Fe_2S_3 or $Cu_2S . FeS_2$	Cu
	Copper Glance	Cu_2S	Cu
	Zinc Blende	ZnS	Zn
	Cinnabar	HgS	Hg
	Galena	PbS	Pb
	Argentite or Silver Glance	Ag_2S	Ag
Carbonate Ores	Magnesite	$MgCO_3$	Mg
	Dolomite	$CaCO_3.MgCO_3$	Mg
	Calamine	$ZnCO_3$	Zn
	Malachite	$CuCO_3.Cu(OH)_2$	Cu
	Limestone	$CaCO_3$	Ca

Halide ores	Rock Salt	NaCl	Na
	Carnallite	$KCl.MgCl_2.6H_2O$	Mg
	Horn Silver	AgCl	Ag
	Sylvine	KCl	K
	Cryolite	$3\ NaF.AlF_3$ or Na_3AlF_6	Al
Sulphate Ores	Epsom Salt	$MgSO_4.7H_2O$	Mg
	Gypsum	$CaSO_4.2H_2O$	Ca
	Barytes	$BaSO_4$	Ba
	Anglesite	$PbSO_4$	Pb
Silicate ores	Asbestos	$CaSiO_3.3MgSiO_3$	Mg
	Felspar	$K_2O.Al_2O_3.6SiO_2$ or $KAlSi_3O_8$	Al
	Mica	$K_2O.3Al_2O_3.6SiO_22H_2O$	Al
Phosphate Ores	Phosphorite	$Ca_3(PO_4)_2$	P

2.2 Sources from earth, living systems and in sea

Sources from earth

Nearly 80 elements are obtained from mineral deposits on or beneath the surface of the earth.

(a) Metals which are sufficiently unreactive to occur uncombined (i.e. in elementary form) are present in group 10 and 11 of the 2nd and 3rd transition series (e.g Pt, Au, Ag; free Ag is also found in nature).

(b) Metalloids (e.g Ge, As, Sb) and neighbouring metals, all of which have relatively large ionization energies, generally occur as sulphides.

(c) The more strongly metallic elements that form positive ions readily are found as oxides (transition metals), carbonates (group 2 metals) or chlorides (group 1 metals).

(i) Three noble metals (Cu, Ag, Au), Hg and six platinum metals (Ru, Os, Rh, Ir, Pd and Pt) occur in nature in free state. All other metals are found in combined state in the nature.

(ii) The composition of earth's crust is: O(49.1%), Si(26%), Al(7.5%), Fe(4.2%), Ca(3.2%), Na(2.4%), Mg(2.3%) and H (1.0%) by weight.

(iii)In combined state metals are found as (a) Oxides- Mg, Cu, Zn, Al, Mn, Fe, etc. (b)Carbonates-Na, Cu, Mg, Ca, Ba, Zn, Fe etc.(c) Phosphates- Ca, Pb, Fe etc (d) Silicates- Li, Cu, Zn, Al, Mn, Ni, etc and (e) Sulphates- Fe, Cu, Pb, Hg etc.

Source from sea

Four elements such as Na, Mg, Cl_2 and Br_2 can be extracted from the oceans or salt brines, where they are present as monoatomic ions (Na^+, Mg^{2+}, Cl^-, Br^-).

Source from living system

Table 3.2 represents the descending mass abundance of elements. About 30 percent of enzymes have a metal atom at the active site. A large number of biomolecules contain metal ions; many of these molecules are proteins. In addition metal ions in the form of crystalline minerals or amorphous solids are important as structural materials in many organisms.

Table 2.2 Descending mass abundance of the elements

Earth Crust	Oceans	Humans
O	O	H
Si	H	O
Al	Cl	C
Fe	Na	N
Ca	Mg	Na
Mg	S	K
Na	Ca	Ca
K	K	Mg
Ti	C	P
H	Br	S
P	B	Cl

2.3 Purification of ores

The ore is generally associated with rock impurities like clay, sand etc. called '**gangue or matrix**'. The purification of ore means removal of gangue from the powdered ore. This process is otherwise called concentration of the ore or ore dressing. Thus, the percentage of the ore in the concentrated ore is higher than that in the original ore. The purification or concentration of ore can be brought about in the following ways, depending upon the type of ore.

(i)Gravity separation process or hydraulic washing

This method is especially suitable for heavy 'oxide' ores like haematite, tinstone, etc. In this, the powdered ore is placed on a sloping floor (or platform) and washed by directing on it a strong current of water. The lighter sandy, and earthy impurities are washed away; while the heavier ore particles are left behind.

(ii)Froth flotation process

This method is especially suitable for sulphide ores like zinc blende (ZnS), and copper pyrites ($CuFeS_2$). This process is based on the fact that the sulphide ore particles are only moistened by **oil;** while those of oxide, and gangue particles are moistened only by **water**. In this process, the powdered ore is mixed with water and a little pine oil (a foaming agent) and the whole mixture is then stirred vigorously by blowing compressed air. The oil forms a foam (or froth) with air. The ore particles stick to the froth, which rises to the surface; while the rocky, and earthy impurities (gangue) are left in water Fig. 2.1. The froth is skimmed off, collected, and allowed to subside to get concentrated ore.

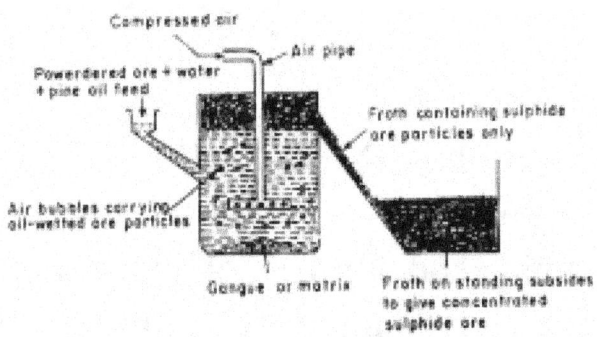

Fig. 2.1 Froth flotation process

(iii)Electromagnetic separation process

This method is meant for separating **magnetic impurities from non-magnetic ore particles,** e.g., tinstone (a tin ore) in which tinstone isnon-magnetic; while impurities iron, manganese and tungstates are magnetic. The powdered ore (containing the associated magnetic impurities) is made to fall (from a hopper) on a belt moving over electromagnetic roller. The magnetic impurities fall from the belt in a heap near the magnet, due to attraction; while the non-magnetic concentrated ore falls in separate heap, away from the magnet, due to the influence of centrifugal force Fig. 2.2.

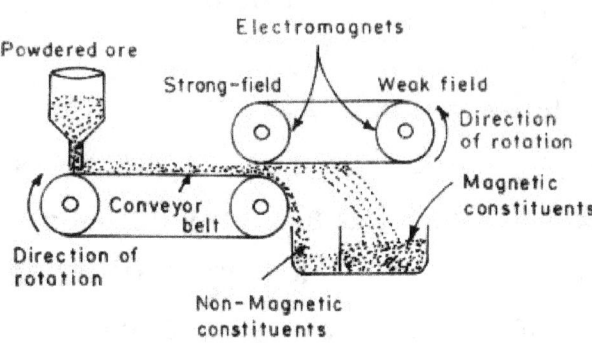

Fig. 2.2 Electromagnetic separation method

(iv)Chemical method

This method is employed in case where the ore is to be in a very pureform, e.g., aluminium extraction. Bauxite (Al_2O_3), an ore of aluminium, contains SiO_2 and Fe_2O_3 as impurities. When bauxite ore is treated with NaOH, the Al_2O_3 goes into solution as sodium meta-aluminate leaving behind the undissolved impurities [Fe_2O_3, SiO_2, $Fe(OH)_3$,etc.], which are then filtered off.

$$Al_2O_3 + 2NaOH \rightarrow 2NaAlO_2 + H_2O$$

Sod. meta. aluminate (In solution form)

The filtrate (containing sodium meta-aluminate) on dilution, and stirring gives a precipitate of aluminium hydroxide, which is filtered, and ignited to get pure alumina.

$$NaAlO_2 + 2H_2O \rightarrow Al(OH)_3 + NaOH$$
Ppt
$$2Al(OH)_3 \rightarrow Al_2O_3 + 3H_2O$$
Pure

2.4 Metallurgical processes

Metallurgy is a branch of chemistry which deals with,

(i)   Extraction of metals from ores

(ii)   Refining of crude metal

(iii) Producing alloys and the study of their constitution, structure and properties.

(iv)   The relationship of physical and mechanical treatment of metals to alloys.

The extraction of metals cannot be carried out by any universal method because extraction of each metal requires different methods of extraction. This depends upon the nature and preparation of metals. In general, noble metals such as Au, Ag, etc are usually extracted by electrolysis of their chlorides, oxides or hydroxides. Heavy metals, e.g. Cu, Zn, Fe, Pb, Sn, etc., are extracted by making use of roasting and smelting methods.

2.4.1 Roasting- oxidation

Roasting is one of the oxidation method where ore is converted into metal oxide. In the process of roasting, the ore either alone or with the addition of suitable material, is subjected to the action of heat in excess of air at temperature below its melting point. Roasting is usually carried out in a reverberatory furnace or in a blast furnace. During roasting

(a) Volatile impurities like S, As, Sb etc. get oxidized and escape out as volatile gases SO_2, As_2O_3 and Sb_2O_3 (b) The sulphide ores decompose to their oxides evolving SO_2 (c) The moisture is removed. Mass becomes porous and thus it can easily be reduced. Roasting may be of many types.

Oxidizing Roasting – In this type of roasting S, As, and Sb impuritiesare removed in the form of their volatile oxides as SO_2, As_2O_3 and Sb_2O_3 etc. due to combined action of heat and air. The ore is simultaneously converted into its oxides. This type of roasting is carried out for ores like copper pyrites, zinc blende and lead ores (PbS) etc.

$$2ZnS + 3O_2 \rightarrow 2ZnO + 2SO_2$$
$$2PbS + 3O_2 \rightarrow 2PbO + 2SO_2$$

Calcination

Another method of conversion of ore into metal oxide (oxidation) is called calcination. It is the process in which the ore is subjected to the action of heat at high temperature in the absence of air below its melting point. The process of calcination is carried out in the case of carbonate and hydrated ore. As a result of calcination (a) The moisture is removed.(b) Gases may be expelled. (c) Volatile impurities are removed. (d) The mass becomes porous. (e) Thermal decomposition of the ore takes place. For example,

$CaCO_3$ (limestone) $\rightarrow CaO + CO_2$

$MgCO_3$ (Magnetite) $\rightarrow MgO + CO_2$

$MgCO_3.CaCO_3$ (Dolomite) $\rightarrow$ MgO + CaO + $2CO_2CuCO_3.Cu(OH)_2$ (Malachite) $\rightarrow 2CuO + H_2O + CO_2$

$ZnCO_3$ (Calamine) $\rightarrow ZnO + CO_2$

$2Fe_2O_3.3H_2O$ (Limonite) $\rightarrow 2Fe_2O_3 + 3H_2O$

The name calcination originated from the ore calcite which on thermal decomposition gives quick lime. Calcination is usually carried out in reverberatory furnace.

2.4.2 Smelting – Reduction

Smelting is one of reduction method where the metal oxide is converted into metal. It is the process used for all operations where the metal is separated by fusion from the ore. The process of smelting is that in which ore is melted with a flux and often with a reducing agent, and it involves, calcination, roasting and reduction. In general, the process of separation of a metal or its sulphide mixture from its ore in fused state is called smelting. Smelting is generally carried out in a blast furnace and high temperature is produced by burning coal or by using electric energy.

In smelting, the roasted or calcined ore is mixed with coke and then heated in a furnace. As a result, (carbon and CO produced by the incomplete combustion) carbon reduces the oxide to the metal. For example, in the extraction of iron, haematite ore(Fe_2O_3) is smelted with coke and limestone (flux). As a result of reduction, iron is obtained in fused or molten state.

$$Fe_2O_3 + 3C \rightarrow 2Fe + 3CO$$
$$CaCO_3 \rightarrow CaO + CO_2$$
$$Fe_2O_3 + 3CO \rightarrow 2Fe + 3CO_2$$
$$CaO + SiO_2 \rightarrow CaSiO_3$$
$$Flux + Gangue \rightarrow Slag$$

Similarly, in the extraction of copper from **copper pyrites,** the ore is mixed with coke and heated in blast furnace (smelted). Infusible impurity FeO is converted to $FeSiO_3$(slag) and is removed. A mixture containing sulphide of copper and iron, called **matte** is formed in the molten state.

$$FeO + SiO_2 \rightarrow FeSiO_2$$

Gangue Flux   Slag

Other Examples

$$ZnO + C \rightarrow Zn + CO$$
$$SnO_2 + 2C \rightarrow Sn + 2CO$$
$$MnO_2 + 2C \rightarrow Mn + 2CO$$

2.4.3 Bessemerization

It is the process used for the manufacture of steel from pig iron. Steel is an alloy of carbon and iron and contains 0.15-1.5% of carbon with traces of sulphur, phosphorus, manganese and silicon as impurities. Depending upon the carbon content, steel are classified into three classes namely mild carbon steel (0.15-0.3%), medium carbon steel (0.3-0.8%) and high carbon steel (0.8-1.50).

The process was discovered by Henry Bessemer in England (1856). The principle involved in this process is that cold air blowed through refractory lined vessel known as converter containing molten pig iron at about 2 atmospheric pressure, oxidizing the impurities and simultaneously converting pig iron to steel.

This process mainly differs in the use of acidic and basic

refractory linings of the converters. In this process low phosphorus pig iron (below 0.09%) is treated by acidic Bessemer process and high phosphorus pig iron (more than 1.5%) is treated in basic.

The converter is a pear shaped furnace about 6m high and 3m in diameter. It is made of steel plates and is lined inside with silica or magnesia (MgO), depending upon the nature of impurities present in the pig iron. If the impurities present in the pig iron are basic, e.g., manganese, a lining of silica brick is used and the process is known as **acid Bessemer process**. If impurities are acidic, e.g., sulphur,phosphorus etc., a basic lining of lime (CaO) or magnesia (MgO) is used in the converter and process is then known as **basic Bessemer process.** The capacity of the converter is from 10-25 tonnes of charge at a time. The converter is mounted on shafts or trunnions, one of which is hollow and serves as a wind pipe and upon which the converter can rotate in any position. The converter is also provided with a number of holes at the bottom through which a hot blast of air can be introduced.

The molten pig iron is mixed in mixers and then charged into converter. About 15-16 tonnes of iron can be charged at a time. The converter is first set in the horizontal position and after charging the converter is adjusted in vertical position. After charging a blast of cold air is admitted through the hole provided at the bottom at a pressure of about 2-3 kg/cm^3. The blast is continued for about 15 minutes during which the impurities are oxidized. Mn is oxidized to MnO and Si is oxidized to SiO_2. Carbon is also oxidized to CO. The resulting oxides of Mn and Si (MnO and SiO_2) combine together to form slag of manganese silicate Fig.3.4.1.

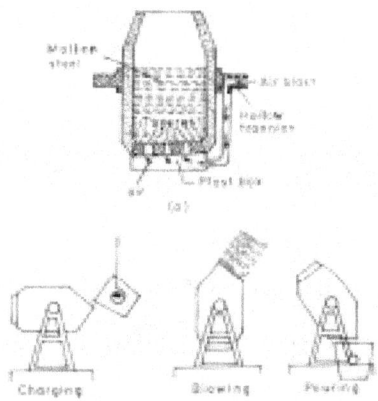

Fig. 2.3 Bessemer Converter and Bessemer Process

2.4.4 Purification of metals
(a) Electrolytic refining

This is one of the most convenient and important method of refining and gives a metal of high purity. This method is applicable to many metals such as **Cu, Ag, Pb, Au, Ni, Sn, Zn etc.** The blocks of impure metal form the anode and thin sheets of pure metal form the cathode. A solution of a salt of the metal is taken as an electrolyte. On passing an electric current through the solution pure metal dissolves from the anode and deposits on cathode. By this process, more metal ions undergo reduction and pure metal is deposited at the cathode. The insoluble impurities either dissolve in the electrolyte or fall at the bottom and collect as **anode mud**. For example, in the refining of copper, impurities like Fe and Zn dissolve in the electrolyte, while Au, Ag and Pt are left behind as anode mud.

(i) **Copper:** During the electrolytic refining of a copper, athick block of impure copper is made anode, and thin plate of pure copper is made cathode. Copper sulphate solution is used as an electrolyte. On passing electric current, following reactions take place:

1) Cu^{2+} ions (from copper sulphate solution) go to the cathode (negative electrode), where they are reduced to copper, which gets deposited on the cathode.

$$Cu^{2+} + 2e^- \rightarrow Cu$$

2) Copper (of impure anode) forms copper ions, and these go into

solution of electrolyte.

$$Cu \rightarrow Cu^{2+} + 2e^-$$

Thus, the net result is transfer of pure copper from anode to the cathode. Impurities like zinc, iron, etc., go into solution; while noble impurities like silver, gold, etc., are left behind as anode mud. Copper is refined to 99.98% pure copper by electrolytic refining (Fig. 2.4).

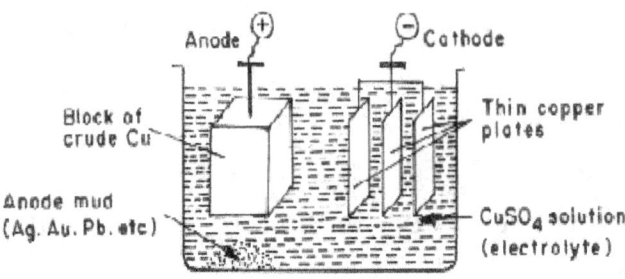

Fig. 2.4 Electrolytic refining of copper

(b) Zone refining

This method is employed for preparing highly pure metal (such as silicon, tellurium, germanium), which are used as semiconductors. It is based on the principle that melting point of a substance is lowered by the presence of impurities. Consequently, when an impure molten metal is cooled, crystals of the pure metal are solidified, and the impurities remain behind the remaining metal. (Fig. 2.5).

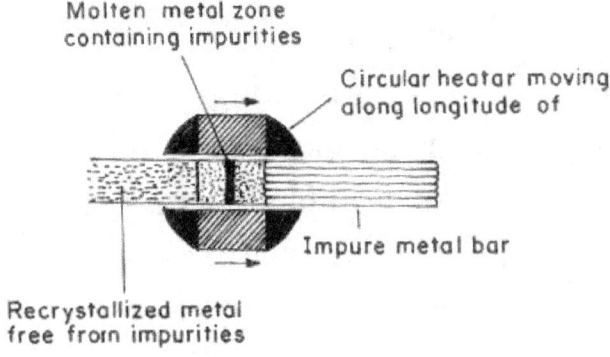

Fig. 2.5 Zone refining

The process consists in casting the impure metal in the form of a bar. A circular heater fitted around this bar is slowly moved longitudinally from one end to the other. At the heated zone, the bar melts, and as the heater moves on, pure metal crystallizes, while the impurities pass into the adjacent molten part. In this way, the impurities are swept from one end of the bar to the other. By repeating the process, ultra pure metal can be obtained.

(c) Mond's process

Thermal methods include methods as carbonyl method, decomposition of hydrides etc. The carbonyl method is used for the refining of metals like Ni and Fe. For example, in case of nickel, the impure metal is heated with CO. The nickel carbonyl thus formed is then decomposed (after distilling off the impurities) to get pure nickel metal and CO. The process is known as **Mond's process.**

$$Ni + 4CO \rightarrow Ni(CO)_4 \rightarrow Ni + 4CO$$

It is based on the following facts:

(a) Only nickel (and not Cu, Fe, etc.) forms a volatile carbonyl, $Ni(CO)_4$, when CO is passed over it at 50^0C.

(b) the nickel carbonyl decomposes at 180^0C to yield pure nickel.

SUMMARY

· This chapter explains and summarizes the salient features of metallurgy

· Different types of ores and their purification methods

· Chemistry behind roasting, smelting and bessemerisation processes

· Various methods of refining process

REFERENCES:

1. Engineering chemistry - P.C.Jain and Monika Jain Twelth Ed - 1999.

2. Industrial chemistry - B.K.Sharma, Tenth Ed. - 1999.

3. A text book of Material Science and metallurgy - O.P.Khanne - 1996.

4. Basic Resource Atlas of Tamil Nadu.

CHAPTER – 3

ATOMIC STRUCTURE

OBJECTIVES
After Studying this Chapter you will able to:
- *Know about the history of structure of atom.*
- *Recognise the merits and demerits of Niels Bohr's model of an atom.*
- *Learn about Sommerfield's atom model.*
- *Analyse the significance of quantum number.*
- *Learn about the shapes of orbitals.*
- *Understand the quantum designation of electron.*
- *Know the application of Pauli's exclusion principle, Hund's rule and Aufbau principle.*
- *Understand the stability of different orbitals and its application in writing the electronic configuration.*
- *Know about the classification of elements based on electronic configuration.*

3.1 Brief introduction of history of structure of atom

The introduction of the atomic theory by John Dalton in 1808 marks the inception of a modern era in chemical thinking. According to this theory, all matter is composed of very small particles called atoms. The atoms were regarded to be structure less, hard, impenetrable particles which cannot be subdivided. Dalton's ideas of the structure of matter were born out by a considerable amount of subsequent experimental evidences towards the end of the nineteenth century. Early in the twentieth century, it has been proved that an atom consists of smaller particles such as electrons, protons and neutrons. The proton, a positively charged particle, is present in the central part of the atom called nucleus. The electron, a negatively charged particle, is present around the nucleus. The neutron, a neutral particle, is also present in the nucleus of the atom. Since the atom is electrically neutral, the number of positive charges on the nucleus is exactly balanced by an equal number of orbital electrons.

Thomson's Model of atom

In 1904 Sir J. J. Thomson proposed the first definite theory as to the internal structure of the atom. According to this theory the atom was assumed to consist of a sphere of uniform distribution of about 10^{-10}m positive charge with electrons embedded in it such that the number of electrons equal to the number of positive charges and the atom as a whole is electrically neutral.

This model of atom could account the electrical neutrality of atom, but it could not explain the results of gold foil scattering experiment carried out by Rutherford.

Rutherford's Scattering Experiment

Rutherford conducted a scattering experiment in 1911 to find out the arrangement of electrons and protons. He bombarded a thin gold foil with D VWUHDP RI IDVW PRYLQJ SRVLWLYHO\ FKDUJHG .-particles emanating from radium.

Rutherford's Nuclear model of a atom

This model resulted from conclusion drawn from experiments on the scattering of alpha particles from a radio active source when the particles were passed through thin sheets of metal foil. According to him

(i) 0RVW RI WKH VSDFH LQ WKH DWRP LV HPSW\ DV PRVW RI WKH .-particles passed through the foil.

(ii) $ IHZ SRVLWLYHO\ FKDUJHG .-particles are deflected. The deflection must be due to enormous repulsive force showing that the positive charge of the atom is not spread throughout the atom as Thomson had thought. The positive charge has to be concentrated in a very smallYROXPH WKDW UHSHOOHG DQG GHIOHFWHG WKH SRVLWLYHO\ FKDUJHG.-particles. This very small portion of the atom was called

nucleusby Rutherford.

(iii) Calculations by Rutherford showed that the volume occupied by the nucleus is negligibly small as compared to the total volume of the atom. The diameter of the atom is about 10^{-10} m while that of nucleus is 10^{-15}m. One can appreciate this difference in size by realizing that if a cricket ball represents a nucleus, then radius of the atom would be about 5 km.

On the basis of above observations and conclusions, Rutherford proposed the nuclear model of atom. According to this model:

(a) An atom consists of a tiny positively charged nucleus at its centre.

(b) The positive charge of the nucleus is due to protons. The mass of the nucleus, on the other hand, is due to protons and some neutral particles each having mass nearly equal to the mass of proton. This neutral particle, called neutron, was discovered later on by Chadwick in 1932. Protons and neutrons present in the nucleus are collectively also known as nucleons. The total number of nucleons is termed as **mass number(A)** of the atom.

(c) The nucleus is surrounded by electrons that move around the nucleus with very high speed in circular paths called **orbits**. Thus, Rutherford's model of atom resembles the solar system in which the sun plays the role of the nucleus and the planets that of revolving electrons.

(d) The number of electrons in an atom is equal to the number of protons in it. Thus, the total positive charge of the nucleus exactly balances the total negative charge in the atom making it electrically neutral. The number of protons in an atom is called its atomic number(**Z**).

(e) Electrons and the nucleus are held together by electrostatic forces of attraction.

3.2 Defects of Rutherford's model

According to Rutherford's model, an atom consists of a positive nucleus with the electrons moving around it in circular orbits. However it had been shown by J. C. Maxwell that whenever an electron is subjected to acceleration, it emits radiation and loses energy. As a result of this, its orbit should become smaller and smaller Fig. 3.1. and finally it should drop into the nucleus by following a spiral path. This means that atom would collapse and thus Rutherford's model failed to explain stability of atoms.

Another drawback of the Rutherford's model is that it says nothing about the electronic structure of the atoms i.e., how the electrons are distributed around the nucleus and what are the energies of these electrons. Therefore, this model failed to explain the existence of certain definite lines in the hydrogen spectrum.

Postulates of Bohr's model of an atom

To overcome the above defects of Rutherford's model, Niels Bohr in 1913 gave a modification based on Quantum theory of radiation. The important postulates are:

(1) The electrons revolve round the nucleus only in certain selected circular paths called orbits. These orbits are associated with definite energies and are called energy shells or energy levels or quantum levels. These are numbered as 1, 2, 3, 4 ….. etc. (starting from the nucleus) are designated as K, L, M, N ….etc. (Fig. 3.2).

(2) As long as an electron remains in a particular orbit, it does not lose or gain energy. This means that energy of an electron in a particular path remains constant. Therefore, these orbits are also called stationary states.

(3) Only those orbits are permitted in which angular momentum of the HOHFWURQ LV D ZKROH QXPEHU PXOWLSOH RI K i ZKHUH μK¶ LV 3ODQFN¶V constant. An electron moving in a circular orbit has an angular momentum equal to mvr where m is the mass of the electron and v, the angular momentum, mvU LV D ZKROH QXPEHU PXOWLSOH RI K i L H

$$mvU \quad QK \; i \; ZKHUH \; Q \quad \text{«««}$$

In other words, angular velocity of electrons in an atom is quantized.

(4) If an electron jumps from one stationary state to another, it will absorb or emit radiation of a definite frequency giving a spectral line of that frequency which depends upon the initial and final levels. When an electron jumps back to the lower energy level, it radiates same amount of energy in the form of radiation.

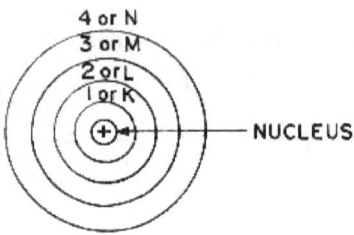

Fig. 3.2 Bohr's orbits

Limitation of Bohr's Theory

(i) According to Bohr, the radiation results when an electron jumps from one energy orbit to another energy orbit, but how this radiation occurs is not explained by Bohr.

(ii) Bohr Theory had explained the existence of various lines in H-spectrum, but it predicted that only a series of lines exist. At that time this was exactly what had been observed. However, as better instruments and techniques were developed, it was realized that the spectral line that had been thought to be a single line was actually a collection of several lines very close together (known as fine spectrum). Thus for example, the single Hv-spectral line of Balmer series consists of many lines very close to each other.

(iii) Thus the appearance of the several lines implies that there are several energy levels, which are close together for each quantum number n. This would require the existence of new quantum numbers.

(iv) Bohr's theory has successfully explained the observed spectra for hydrogen atom and hydrogen like ions (e.g. He^+, Li^{2+}, Be^{3+} etc.), it cannot explain the spectral series for the atoms having a large number of electrons.

(v) There was no satisfactory justification for the assumption that the electron can rotate only in those orbits in which the angular momentum of the electron (mvU LV D ZKROH QXPEHU PXOWLSOH RI K i i.e. he could not give any explanation for using the principle of quantization of angular momentum and it was introduced by him arbitrarily.

(vi) Bohr assumes that an electron in an atom is located at a definite distance from the nucleus and is revolving round it with definite velocity, i.e. it is associated with a fixed value of momentum. This is against the Heisenberg's Uncertainty Principle according to which it is impossible to determine simultaneously with certainty the position and the momentum of a particle.

(vii) No explanation for Zeeman effect: If a substance which gives a line emission spectrum, is placed in a magnetic field, the lines of the spectrum get split up into a number of closely spaced lines. This phenomenon is known as Zeeman effect. Bohr's theory has no explanation for this effect.

(viii) No explanation of the Stark effect: If a substance which gives a line emission spectrum is placed in an external electric field, its lines get spilt into a number of closely spaced lines. This phenomenon is known as Stark effect. Bohr's theory is not able to explain this observation as well.

3.3 Electronic configuration and quantum numbers
Quantum Numbers

The quantum numbers are nothing but the details that are required to locate an electron in an atom. In an atom a large number of electron orbitals are permissible. An orbital of smaller size means there is more chance of finding the electron near the nucleus. These orbitals are designated by a set of numbers known as quantum numbers. In order to specify energy, size, shape and orientation of the electron orbital, three quantum numbers are required these are discussed below.

1. The principal quantum number (n)

The electrons inside an atom are arranged in different energy levels called electron shells or orbits. Each shell is characterized by a quantum number called principal quantum number. This is represented by the letter 'n' and 'n' can have values, 1,2,3,4 etc. The first level is also known as K level. Second as L level, third as M level, fourth as N level and so on. The first or K level is the orbit nearest to the nucleus and next one is second or L level and so on.

2. The subsidiary or azimuthal quantum number (*l*)

According to Sommerfield, the electron in any particular energy level could have circular path or a variety of elliptical paths about the nucleus resulting in slight differences in orbital shapes with slightly differing energies due to the differences in the attraction exerted by the nucleus on the electron. This concept gave rise to the idea of the existence of sub-energy levels in each of the principal energy levels of the atom. This is denoted by the letter '*l*' and have values from 0 to n-1.

Thus, if n=1, *l*=0 only one value (one level only) *s* level.

n=2, *l*=0 and 1 (2 values or 2 sub- levels) *s* and *p* level.

n=3, *l*=0, 1 and 2 (3 values or 3 sub-levels) *s, p* and *d* level.

n=4, *l*=0, 1, 2 and 3 (4 values or 4 sub-levels) *s, p ,d* and *f* level.

3. Magnetic quantum number (m)

In a strong magnetic field a sub-shell is resolved into different orientations in space. These orientations called orbitals have slight differences in energy. This explains the appearance of additional lines in atomic spectra produced when atoms emit light in magnetic field. Each orbitals is designated by a magnetic quantum number m and its values depends on the value of '*l*' . The values are -' *l*' through zero to +' *l*' and thus there are (2*l*+1) values.

Thus when *l*=0, m= 0 (only one value or one orbital)

$l=1$, m= -1, 0, +1 (3 values or 3 orbitals)
$l=2$, m= -2, -1, 0, +1, +2 (5 values or 5 orbitals)
$l=3$, m= -3,-2, -1, 0, +1, +2, +3 (7 values or 7 orbitals).

The three quantum numbers labeling an atomic orbital can be used equally well to label electron in the orbital. However, a fourth quantumnumber, the spin quantum number, (s) is necessary to describe an electron completely.

4. Spin quantum number (s)

The electron in the atom rotates not only around the nucleus but also around its own axis and two opposite directions of rotation are possible (clock wise and anticlock wise). Therefore the spin quantum number can have only two values +1/2 or −1/2. For each values of m including zero, there will be two values for s.

To sum up, the four quantum numbers provide the following informations:

1. n identifies the shell, determines the size of the orbital and also to a large extent the energy of the orbit.
2. There are n subshells in the nth shell. l identifies the subshell and determines the shape of the orbital. There are $(2l+1)$ orbitals of each type in a subshell i.e., one s orbital ($l=0$), three p orbitals ($l=1$), and five d orbitals ($l=2$) per subshell. To some extent l also determines the energy of the orbital in a multi-electron atom.
3. m_l designates the orientation of the orbital. For a given value of l, m_l has $(2l+1)$ values, the same as the number of orbitals per subshell. It means that the number of orbitals is equal to the number of ways in which they are oriented.
4. m_s refers to orientation of the spin of the electron.

Example 1

What is the total number of orbitals associated with the principal quantum number n=3 ?

Solution

For n = 3, the possible values of l are 0,1 and 2. Thus, there is one 3s orbital (n = 3, l = 0 and m_l = 0); there are three p orbitals (n = 3, l = 1 and m_l = -1, 0, 1) there are five 3d orbitals (n = 3, l = 2, m_l = -2, -1, 0, 1, 2).

Therefore, the total number of orbitals is 1+3+5 = 9.

Example 2

Using s, p, d, f notations, describe the orbital with the following quantum numbers (a) n=2, l = 1 (b) n = 4, l = 0 (c) n = 5, l = 3 (d) n = 3, l = 2.

Solution

	n	l	Orbital
(a)	2	1	2p
(b)	4	0	4s
(c)	5	3	5f
(d)	3	2	3d

3.4 Shapes or boundary surfaces of Orbitals

s-orbitals: For s-orbital l = 0 and hence, m can have only one value, i.e., m = 0. This means that the probability of finding the electron in s-orbital is the same in all directions at a particular distance. In other words s-orbitals are spherically symmetrical.

The electron cloud picture of 1s-orbital is spherical. The *s*-orbitals of higher energy levels are also spherically symmetrical. However, they are more diffused and have spherical shells within them where probability of finding the electron is zero. These are called nodes. In 2*s*-orbital there is one spherical node. In the ns orbital, number of nodes are (n-1).

p-orbitals: For p-orbitals l = 1 and hence 'm' can have three possible values +1, 0, -1. This means that there are three possible orientations of electron cloud in a *p*-sub-shell. The three orbitals of a *p*-sub-shell are designated as p_x, p_y and p_z respectively along x-axis, y-axis and z-axis respectively. Each *p*-orbital has two lobes, which are separated by a point of zero probability called node. Each *p*-orbital is thus dumb bell shaped.

In the absence of magnetic field these three p-orbitals are equivalent in energy and are, therefore, said to be three-fold degenerate or triply degenerate. In the presence of an external magnetic field, the relative energies of the three p orbitals vary and depend on their orientation ormagnetic quantum number. This probably accounts for the splitting of a single spectral line into a number of closely spaced lines in presence of a magnetic field (fine structure).

***d*-orbitals:**For$d$-orbitals$l$= 2, m = 0, ±1, ±2 indicating that d-orbitals have five orientations, i.e., there are five d-orbitals which are named as d_{xy}, d_{yz}, d_{zx}, d_z^2 and $d_{x^2-y^2}$. All these five orbitals, in the absence of magnetic field, are equivalent in energy and are, therefore, said to be five-fold degenerate

The three orbitals namely d_{xy}, d_{yz} and d_{zx} have their lobes lying symmetrically between the coordinate axes indicated in the subscript to d, e.g. the lobes of d_{xy} orbital are lying between the x-and y-axes. This set of three orbitals is known as t_{2g} set. $d_{x^2-y^2}$ and d_z^2 orbitals have their lobes along the axes (i.e. along the axial directions), e.g., the lobes of d orbital lie along the x and y-axes, while those of d_z^2 orbital lie along the z-axis. This set is known as e_g set.

3.5 Pauli's exclusion principle

The filling of electron into the orbitals of different atoms takes place according to the Aufbau principle, which is based on the Pauli's exclusion principle and the Hund's rule of maximum multiplicity.

The distribution of quantum numbers among the electrons in a given atom is governed by Pauli's Exclusion principle, which states that ***"it isimpossible for any two electrons in a given atom to have all the four quantum numbers identical"*** i.e., in an atom, two electrons can havemaximum three quantum numbers (n, l and m) the same and the fourth (s) will definitely be having a different value. Thus if s = +1/2 for one electron, s should be equal to −1/2 for the other electron. In other words the two electrons in the same orbital should have opposite spins.

Uses of the principle

The greatest use of the principle is that it is helpful in determining the maximum number of electrons that a main energy level can have. Let us illustrate this point by considering K and L shells.

(a) K-shell: For this shell $n = 1$. For $n = 1$, $l = 0$ and $m = 0$. Hence s can have a value either $+1/2$ or $-1/2$. The different values of n, l, m and s given above give the following two combinations of the four quantum numbers, keeping in view the exclusion principle. Combination (i) is for one electron and combination (ii) is for the other electron.

(i) $n = 1$, $l = 0$, $m = 0$
$s = +1/2$ (1st electron)

(ii) $n = 1$, $l = 0$, $m = 0$,
$s = -1/2$ (2nd electron)

(Two electrons in $l = 0$ sub-shell i.e., 1s-orbital)

These two combinations show that in K shell there is only one sub-shell corresponding to $l = 0$ value (s-sub-shell) contains only two electrons with opposite spins.

(b) L-shell: For this shell $n = 2$. For $n = 2$ the different values of l, m and s give the following eight combinations of four quantum numbers.

(i) $n = 2$, $l = 0$, $m = 0$, $s = +1/2$
(ii) $n = 2$, $l = 0$, $m = 0$, $s = -1/2$
(iii) $n = 2$, $l = 1$, $m = 0$, $s = +1/2$
(iv) $n = 2$, $l = 1$, $m = 0$, $s = -1/2$
(v) $n = 2$, $l = 1$, $m = +1$, $s = +1/2$
(vi) $n = 2$, $l = 1$, $m = +1$, $s = -1/2$
(vii) $n = 2$, $l = 1$, $m = -1$, $s = +1/2$
(viii) $n = 2$, $l = 1$, $m = -1$, $s = -1/2$

Eight combinations given above show that L shell is divided into two sub-shells corresponding to $l = 0$ (s sub-shell) and $l = 1$ (p sub-shell) and this shell cannot contain more than 8 electrons, i.e., its maximum capacity for keeping the electrons is eight.

3.6 Hund's rule of maximum multiplicity

Hund's rule of maximum multiplicity states, that in filling p, d or f orbitals, as many unpaired electrons as possible are placed before pairing of electrons with opposite spin is allowed. Pairing of electrons requires energy. Therefore **no pairing occurs until all orbitals of a given sub-level are half filled.** This is known as Hund's rule of maximummultiplicity. It states that when electrons enter sub-levels of fixed (n+1) values, available orbitals are singly occupied (Table 3.1).

Table 3.1 Representation of arrangements of electrons

Atomic Number	Element	1s	2s	$2p_x$	$2p_y$	$2p_z$	Number of unpaired electrons
1	H	9					1
2	He	9;					0
3	Li	9;	9				1
4	Be	9;	9;				0
5	B	9;	9;	9	9		1
6	C	9;	9;	9	9		2
7	N	9;	9;	9	9	9	3
8	O	9;	9;	9;	9	9	2
9	F	9;	9;	9;	9;	9	1
10	Ne	9;	9;	9;	9;	9;	0

Thus, if three electrons are to be filled in the p- level of any shell, one each will go into each of the three (p_x, p_y, p_z) orbitals. The fourth electron entering the p- level will go to p_x orbital which now will have two electrons with opposite spins (as shown above) and said to be paired. The unpaired electrons play an important part in the formation of bonds.

3.7 Aufbau Principle

The word 'aufbau' in German means 'building up'. The building up of orbitals means the filling up of orbitals with electrons. The principles states: **In the ground state of the atoms, the orbitals are**

filled in order of their increasing energies. In other words, electrons firstoccupy the lowest-energy orbital available to them and enter into higher energy orbitals only after the lower energy orbitals are filled. The order in which the energies of the orbitals increase and hence the order in which the orbitals are filled is as follows:

1s, 2s, 2p, 3s, 3p, 4s, 3d, 4p, 5s, 4d, 5p, 6s, 4f, 5d, 6p, 7s...............

This order may be remembered by using the method given in Fig. 3.3. Starting from the top, the direction of the arrows gives the order of filling of orbitals. Alternatively, the order of increase of energy of orbitals can be calculated from (n +1) rule, explained below:

The lower the value of (n+1) for an orbital, the lower is its energy. If two orbitals have the same (n+1) value, the orbital with lower value of n has the lower energy.

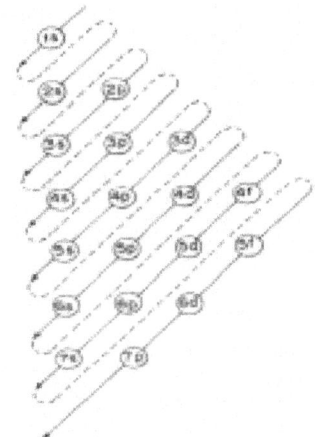

Fig. 3.3 Order of filling of orbitals

It may be noted that different subshells of a particular shell have different energies in case of many-electron atoms. However in hydrogen atom, they have the same energy.

3.8 Stability of orbitals

According to Hund's rule atoms having half-filled or completely-filled orbitals are comparatively more stable and hence more energy is needed to remove an electron from such atoms. The ionization potential or ionization enthalpy of such atom is, therefore, relatively higher than expected normally from their position in the periodic table.

The extraordinary stability of half-filled and completely filled electron configuration can be explained in terms of symmetry and exchange energy. The half-filled and completely filled electron configurations have symmetrical distribution of electrons and this symmetry leads to stability. Moreover, in such configuration electron can exchange their positions among themselves to maximum extent. This exchange leads to stabilization for example, half-filled $2p$ orbital is Nitrogen and completely filled orbitals in Neon are given as follows.

Thus the $p^3, p^6, d^5, d^{10}, f^7$ and f^{14} configuration which are either completely filled or exactly half filled are more stable.

Further, it may be noted that chromium and copper have five and ten electrons in $3d$ orbitals rather than four and nine electrons respectively as expected. Therefore, to acquire more stability one of the 4s electron goes into $3d$ orbitals so that $3d$ orbitals get half-filled or completely filled in chromium and copper respectively.

Chromium

Expected configuration: $1s^2, 2s^2, 2p^6, 3s^2, 3p^6, 3d^4, 4s^2$

Actual configuration: $1s^2, 2s^2, 2p^6, 3s^2, 3p^6, 3d^5, 4s^1$

Electron exchange

Copper

Expected configuration: $1s^2, 2s^2, 2p^6, 3s^2, 3p^6, 3d^9, 4s^2$

Actual configuration: $1s^2, 2s^2, 2p^6, 3s^2, 3p^6, 3d^{10}, 4s^1$

Electron exchange

SUMMARY

· The model of the nuclear atom developed by Thomson, Rutherford and their defects are explained.

· Niel's Bohr model and sommerfield's extension were mentioned with diagrammatic representation.

· The location of electron in an atom through four quantum numbers are explained including their significance.

· Occupancy of electrons following Hund's rule Aufbau principle,

Pauli's exclusion principle are explained with represnetation.

REFERENCES:

1) Concise Inorganic chemistry - J.D. Lee, 3rd Edn - 1977 and 5th Edn-2002.

2) Theoretical Inorganic chemistry - M.C.Day and J.Sellbin, 2nd Edn. - 1985.

3) Theorical princples of Inorganic chemistry - G.S.Manker, 9th Edn - 1993.

4) Selected topics in Inorganic Chemistry - U.Mallik, G.D.Tute and R.D. Madan, 6th Edn. - 1993c.

CHAPTER – 4

PERIODIC CLASSIFICATION

OBJECTIVES

After Studying this Chapter you will able to:

· *Recall the history of periodic classification.*

· *Understand the IUPAC periodic table and the nomenclature of elements with atomic number greater than 100.*

· *Understand the electronic configuration of the elements and the classification based on it.*

· *Analyze the periodicity of properties like atomic radii, ionization potential, electron affinity electronegativity etc.*

· *Know the anomalous periodic properties of elements and to reason it.*

4.1 Brief history of periodic classification

More than one hundred and nine elements are known today. The periodic table of elements is an important landmark in the history of chemistry. It would be difficult to study individually the chemistry of all the elements and their numerous compounds. The periodic table provides a systematic and extremely useful framework for organizing a lot of information available on the chemical behaviour of the elements into a few simple and logical patterns. This gave rise to the necessity of classifying the elements into various groups or families having similar properties. This classification has resulted in the formulation of periodic table. Periodic table may be defined as the arrangements of various elements according to their properties in a tabular form.

All earlier attempts on the classification of elements were based on atomic mass. Several chemists have for long tried to classify the elements and to find patterns in their properties.

Dobereiner's Triads

In 1829, *John Dobereiner* (German Chemist) classified elements having similar properties into groups of three. These groups were called triads. According to this law when elements are arranged in the order of increasing atomic mass in a triad, the atomic mass of the middle element was found to be approximately equal to the arithmetic mean of the other two elements. For example lithium, sodium and potassium constituted one triad. However, only a limited number of elements could be grouped into traids.

Table 4.1 Doberenier's Triads

Elements	Atomic weight	Element	Atomic weight	Element	Atomic weight
Li	7	Ca	40	Cl	35.5
Na	23	Sr	88	Br	80
K	39	Ba	137	I	127

Newlands Law of Octaves

In 1865, John Newlands (English Chemist) observed that if the elements were arranged in order of their increasing atomic weights, the eighth element starting from a given one, possessed properties similar to the first, like the eighth note in an octave of music. He called it the law of octaves. It worked well for the lighter elements but failed when applied to heavier elements.

Lother-Meyer's Arrangement

In 1869, **J. Lother-Meyer** in Germany gave a more detailed and accurate relationship among the elements. Lother-Meyer plotted atomic volumes versus atomic weights of elements and obtained a curve. He pointed out that elements occupying similar positions in the curve possessed similar properties.

Mendeleev's Periodic Table

In 1869, Dimitriv Mendeleev (Russian Chemist) arranged the 63 chemical elements, then known, according to their increasing order of atomic weights. He gave his famous scheme of the periodic classification of elements known as the periodic law. The law states that "The properties of the elements are the periodic function of their atomicweights". It means that when elements are arranged in order of increasing atomic weights, the elements was similar properties recur after regular intervals. On the basis of this periodic law Mendeleev constructed a periodic table in such a way that the elements were arranged horizontally in order of their increasing atomic weights. Mendeleev, while studying his Periodic Table had found that in certain cases the regularity in behaviour between two succeeding elements was not observed. In order to overcome this he had kept gaps between such elements and had predicted that the gaps would be filled by new elements, to be discovered in future, For example, both gallium and germanium were not discovered at the time when Mendeleev proposed the periodic table. Mendeleev named these elements as eka-aluminium and eka-silicon because he believed that they would be similar to aluminium and silicon respectively. These elements were discovered later and Mendeleev's prediction proved remarkably correct. The discoveries/synthesis of new elements have continued even to the present day, raising their number to 120. The elements with atomic numbers upto 92 (i.e. uranium) are found in nature. The rest known as trans uranium elements have been synthesized in the laboratories, which are highly unstable. They decay radioactively.

The modified periodic table is essentially similar to that of Mendeleev with a separate column added for noble gases, which were not discovered until the closing years of the nineteenth century. The general plan of the modified Mendeleev's periodic table is improved.

The Mendeleev's modified periodic table consists of:

(1) Nine vertical columns called groups. These are numbered from I to VIII and zero. (The members of zero group were not

discovered at the time of Mendeleev). Each group from I to VII is further sub-divided into two sub-groups designated as A and B. Group VIII consists of three sets, each one containing three elements. Group zero consists of inert gases.

(2) Seven horizontal rows, called periods. These are numbered from 1 to 7. First period contains two elements. Second and third periods contain eight elements each. These periods are called short periods. Fourth and fifth contains eighteen elements each. These periods are called long periods. Sixth period contains thirty two elements and is called longest period. Seventh period is incomplete and contains nineteen elements according to early classification.

4.2 IUPAC periodic table and IUPAC nomenclature of elements with atomic number greater than 100

Modern Periodic Law

In 1913, a British Physicist Henry Moseley showed that the atomic number is a more fundamental property of an element than its atomic weight. This observation led to the development of modern periodic law. The modern periodic law states that " the physical and chemical properties of the elements are periodic function of their atomic numbers."

This means that when the elements are arranged in order of increasing atomic numbers, the elements with similar properties recur after regular intervals. The periodic repetition is called periodicity. The physical and chemical properties of the elements are related to the arrangement of electrons in the outermost shell. Thus, if the arrangement of electrons in the outermost shell (valence shell) of the atoms is the same, their properties will also be similar. For example, the valence shell configurations of alkali metals show the presence of one electron in the s-orbital of their valence shells.

Similar behaviour of alkali metals is attributed to the similar valence shell configuration of their atoms. Similarly, if we examine the electronic configurations of other elements, we will find that there is repetition of the similar valence shell configuration after certain regular intervals with the gradual increase of atomic number. Thus we find that the periodic repetition of properties is due to the recurrence of similar valence shell configuration after certain intervals. It is observed that similarity in properties is repeated after the intervals of 2, 8, 18, or 32 in their atomic numbers.

Long form of the Periodic Table: The periodic table is constructedon the basis of repeating electronic configurations of the atoms when they are arranged in the order of increasing atomic numbers. The long form of the Periodic table is given in a modified form in page number 70. Readers are advised to follow the periodic table closely while studying the structural features of the long form of the Periodic Table.

Structural Features of the Long form of the periodic Table: Thelong form of the periodic table consists of horizontal rows called periods and vertical columns called groups.

Periods: In terms of electronic structure of the atom, a periodconstitutes a series of elements whose atoms have the same number of electron shell i.e., principal quantum number (n). There are seven periods and each period starts with a different principal quantum number.

The first period corresponds to the filling of electrons in the first energy shell (n = 1). Now this energy level has only one orbital (1s) and, therefore, it can accommodate two electrons. This means that there can be only two elements (hydrogen, $1s^1$ and helium, $1s^2$) in the first period.

The second period starts with the electron beginning to enter the second energy shell (n = 2). Since there are only four orbitals (one 2s- and three 2p- orbitals) to be filled, it can accommodate eight electrons. Thus, second period has eight elements in it. It starts with lithium (Z = 3) in which one electron enters the 2s-orbital. The period ends with neon (Z = 10) in which the second shell is complete $(2s^2 2p^6)$.

The third period begins with the electrons entering the third energy shell (n = 3). It should be noted that out of nine orbitals of this energy level (one s, three p and five d) the five 3d-orbitals have higher energy than 4s-orbitals. As such only four orbitals (one 3s and three 3p) corresponding to n = 3 are filled before fourth energy level begins to be filled. Hence, third period contains only eight elements from sodium (Z = 11) to argon (Z = 18).

The fourth period corresponding to n = 4 involves the filling of one 4s and three 4p-orbitals (4d and 4f orbitals have higher energy than 5s-orbital and are filled later). In between 4s and 4p-orbitals, five 3d-orbitals are also filled which have energies in between these orbitals. Thus, altogether nine orbitals (one 4s, five 3d and three 4p) are to be filled and therefore, there are eighteen elements in fourth period from potassium (Z = 19) to krypton (Z = 36). The elements from scandium (Z = 21) to zinc (Z = 30) are called 3d- transition series.

The fifth period beginning with 5s-orbital (n=5) is similar to fourth period. There are nine orbitals (one 5s, five 4d and three 5p) to be filled and, therefore, there are eighteen elements in fifth period from rubidium (Z = 37) to xenon (Z = 54).

The sixth period starts with the filling of 6s-orbitals (n= 6). There are sixteen orbitals (one 6s, seven 4f, five 5d, and three 6p) in which filling of electrons takes place before the next energy level starts. As such there are thirty two elements in sixth period starting from cesium (Z = 55) and ending with radon (Z = 86). The filling up of 4f orbitals begins with cerium (Z = 58) and ends at lutetium (Z = 71). It constitutes the first f-inner transition series which is called lanthanide series.

The seventh period begins with 7s-orbital (n = 7). It would also have contained 32 elements corresponding to the filling of sixteen orbitals (one 7s, seven 5f, five 6d and three 7p), but it is still incomplete. At present there are 23 elements in it. The filling up of 5f- orbitals begins with thorium (Z = 90) and ends up at lawrencium

(Z = 103). It constitutes second f-inner transition series which is called actinide series. It mostly includes man made radioactive elements. In order to avoid undue extension of the periodic table the 4f and 5f- inner transition elements are placed separately.

The number of elements and the corresponding orbitals being filled are given below.

Table 4.2

Period	Principal Valence shell (=n)	Orbitals being filled Up	Electrons to be accommodated	Number of electrons
First	N = 1	1s	2	2
Second	N = 2	2s, 2p	2+6	8
Third	n = 3	3s, 3p	2+6	8
Fourth	n = 4	4s, 3d, 4p	2 +10+6	18
Fifth	n = 5	5s, 4d, 5p	2+10+6	18
Sixth	n = 6	6s, 4f, 5d, 6p	2+14+10+6	32
Seventh	n = 7	7s, 5f, 6d, 7p	2+14+10+6	32

The first three periods containing 2, 8 and 8 elements respectively are called short periods, the next three periods containing 18, 18 and 32 elements respectively are called long periods.

Groups

A vertical column in the periodic table is known as group. A groupconsists of a series of elements having similar configuration of the outer energy shell. There *are* eighteen vertical columns in long from of the periodic table. According to the recommendations of the **InternationalUnion of Pure and Applied Chemistry** (IUPAC), these groups arenumbered from 1 to 18. Previously, these were numbered from I to VII as A and B, VIII and zero groups elements. The elements belonging to the same group are said to constitute a family. For example, elements of group 17 (VII A) constitute halogen family.

IUPAC Nomenclature for Elements with Z > 100

The elements beyond uranium (Z = 92) are all synthetic elements and are known as transuranium elements. The elements beyond fermium are known as transfermium elements. These elements fermium (Z = 100), mendelevium (Z = 101), nobelium (Z = 102) and lawrencium (Z = 103) are named after the names of famous scientists. Although names and symbols to many of these elements have been assigned by these are still not universally accepted. Also some of these elements have been assigned two names/symbols. For example, element with atomic number 104 is called either Kurchatovium (Ku) or Rutherfordium (Rf) while element with atomic number 107 is called Neilsbohrium (Ns) or Borium (Bh). But the following elements have been assigned only one official name. For example element with atomic number 105 is called Dubnium, with atomic number 106 as Seaborgium, with atomic number 108 as Hassnium and with atomic number 109 is named as Meiternium. To overcome all these difficulties, IUPAC nomenclature has been recommended for all the elements with Z > 100. It was decided by IUPAC that the names of elements beyond atomic number 100 should use Latin words for their numbers. The names of these elements are derived from their numerical roots.

Numerical → Roots	0	1	2	3	4	5	6	7	8	9
	nil	un	bi	tri	quad	pent	hex	sept	oct	en

Table 4.3

Atomic number	Name of the element	Symbol
101	Unnilunnium	Unu
102	Unnilbium	Unb
103	Unniltrium	Unt
104	Unnilquadium	Unq
105	Unnilpentium	Unp
106	Unnilhexium	Unh
107	Unnilseptium	Uns
108	Unniloctium	Uno
109	Unnilennium	Une
110	Ununnilium	Uun
111	Unununium	Uuu

112	Ununbium	Uub
113	Ununtrium	Uut
114	Ununquadium	Uuq
115	Ununpentium	Uup
116	Ununhexium	Uuh
117	Ununseptium	Uus
118	Ununoctium	Uuo
119	Ununennium	Uue
120	Unbinilium	Ubn

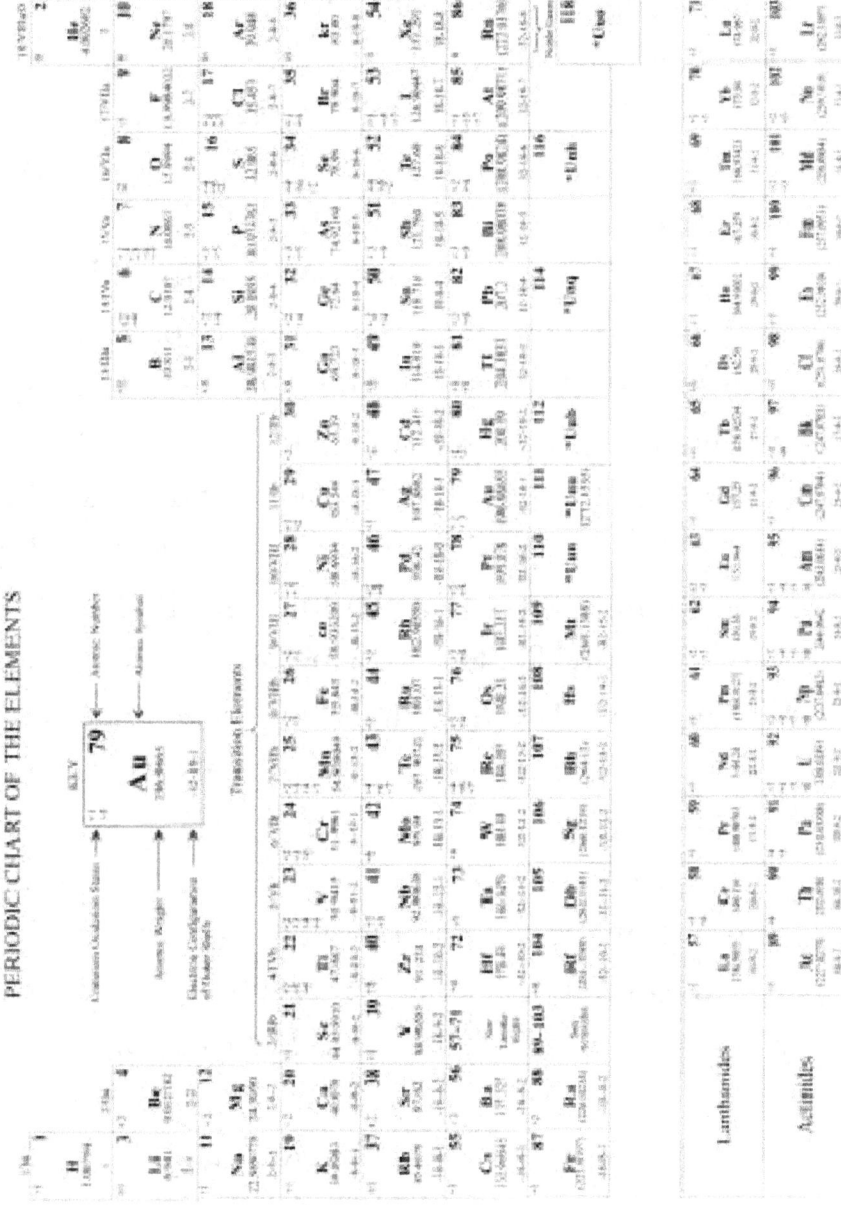

4.3 Electronic configuration and periodic table

There is a close connection between the electronic configuration of the elements and the long form of the Periodic Table. We have already learnt that an electron in an atom is characterized by a set of four quantum numbers and the principal quantum number (n) defines the main energy level known as the **Shell**. The electronic configuration of elements can be best studied in terms of variations in periods and groups of the periodic table.

(a) Electronic Configuration in periods

Each successive period in the periodic table is associated with the filling up of the next higher principal energy level (n=1, n=2,etc.). It can be readily seen that the number of elements in each period is twice the number of atomic orbitals available in the energy level that is being filled. The first period starts with the filling of the lowest level (1s) and has thus the two elements-hydrogen ($1s^1$) and helium ($1s^2$) when the first shell (K) is completed. The second period starts with lithium and the third electron enters the 2s orbital. The next element, beryllium has four electrons and has the electronic configuration $1s^2 2s^2$. Starting from the next element boron, the 2p orbitals are filled with electrons when the L shell is completed at neon ($2s^2 2p^6$). Thus there are 8 elements in the second period. The third period (n=3) begins at sodium, and the added electron enters a 3s orbital. Successive filling of 3s and 3p orbitals gives rise to the third period of 8 elements from sodium to argon.

The fourth period (n=4) starts at potassium with the filling up of 4s orbital. Now you may note that before the 4p orbital is filled, filling up of 3d orbitals becomes energetically favourable and we come across the so-called 3d Transition Series of elements. The fourth period ends at krypton with the filling up of the 4p orbitals. Altogether we have 18 elements in this fourth period. The fifth period (n=5) beginning with rubidium is similar to the fourth period and contains the 4d transition series starting at yttrium (Z=39). This period ends at xenon with filling up of the 5p orbitals. The sixth period (n=6) contains 32 elements and successive electrons enter 6s, 4f, 5d and 6p orbitals, in that order. Filling up of the 4f orbitals begins with cerium (Z=58) and ends at lutetium (Z=71) to give the 4f-inner transition series, which is called the **Lanthanoid Series**. The seventh period (n=7) is similar to the sixthperiod with the successive filling up of the 7s, 5f, 6d and 7p orbitals and includes most of the man-made radioactive elements. This period will end at the element with atomic number 118 which would belong to the noble gas family. Filling up of the 5f orbitals after actinium (Z=89) gives the 5f-inner transition series known as the **Actinoid Series**. The 4f- and 5f- transition series of elements are placed separately in the periodic table to maintain its structure and to preserve the principle of classification by keeping elements with similar properties in a single column.

(b) Groupwise/electronic configuration

Elements in the same vertical column or group have similar electronic configurations, have the same number of electrons in the outer orbitals, and similar properties. Group 1 (the alkali metals) is an example.

Thus it can be seen that the properties of an element have periodic dependence upon the atomic number and not on relative atomic mass (Table 4.4).

Table 4.4 Types of elements: Electronic configuration of alkali metals

Atomic number	Symbol	Electronic configuration
3	Li	$1s^2 2s^1$ or [He] $2s^1$
11	Na	$1s^2 2s^2 2p^6 3s^1$ or [Ne] $3s^1$
19	K	$1s^2 2s^2 2p^6 3s^2 3p^6 4s^1$ or [Ar] $4s^1$
37	Rb	$1s^2 2s^2 2p^6 3s^2 3p^6 4s^2 3d^{10} 4p^6 5s^1$ or [Kr] $5s^1$
55	Cs	$1s^2 2s^2 2p^6 3s^2 3p^6 4s^2 3d^{10} 4p^6 5s^2 4d^{10} 5p^6 6s^1$ or [Xe] $6s^1$

Types of elements: *s-, p-, d-, f-* Blocks

The aufbau principle and the electronic configuration of atoms provide a theoretical foundation for the periodic classification. Theelements in a vertical column of the periodic table constitute a group or family and exhibit similar chemical behaviour. Strictly, helium belongs to the s-block but its positioning in the p-block along with other group 18 elements is justified because it has a completely filled valence shell ($1s^2$) and as a result, exhibits properties characteristic of other noble gases. The other exception is hydrogen. It has a lone s- electron and hence can be placed in group 1 (alkali metals). It can also gain an electron to achieve a noble gas arrangement and hence it can behave similar to a group 17 (halogen family) elements. Because it is a special case, we shall place hydrogen separately at the top of the Periodic Table. We will briefly discuss the salient features of the four types of elements marked in the periodic table.

s-Block Elements

The elements of group 1 (alkali metals) and group 2 (alkaline earth metals) which have ns^1 and ns^2 outermost electronic configuration belong to the **s-block elements.** They are all reactive metals with low ionization enthalpies. They lose the outermost electron(s) readily to form 1+ (in the case of alkali metal) or 2+ ions (in the case of alkaline earth metals). The metallic character and the reactivity increase as we go down the group. The compounds of the s-block elements, with the exception of those of beryllium are predominantly ionic.

p-Block Elements

The **p-Block Elements** comprise those belonging to groups 13 to 18 and together with the s-block elements are called the **RepresentativeElements or Main Group Elements.** The outermost electronicconfiguration varies from ns^2np^1 to ns^2np^6 in each period. Each period ends in a noble gas with a closed shell ns^2np^6

configuration. All the orbitals in the valence shell of the **noble gases** are completely filled by electrons and it is very difficult to alter this stable arrangement by the addition or removal of electrons. The noble gases thus exhibit very low chemical reactivity. Preceding the noble gas family are two chemically important groups of nonmetals. They are the **halogens** (groups 17) and **chalcogens** (group 16). These two groups of elements have highernegative electron gain enthalpies and readily add one or two electrons respectively to attain the stable noble gas configuration. The nonmetalliccharacter increases as we move from left to right across a period and metallic character increases as we go down the group.

The *d*-block Elements (Transition Elements)

These are the elements of group 3 to 12 in the center of the periodic table. These elements are characterized by filling of inner d orbitals by electrons and are therefore referred to as *d*-**Block Elements**. These elements have the outer electronic configuration $(n-1) d^{1-10} ns^{1-2}$. They are all metals. They mostly form colored ions and exhibit variable valency. However, Zn, Cd and Hg, which have the $(n-1)d^{10} ns^2$ electronic configuration, do not show most of the properties of transition elements in a way, transition metals form a bridge between the chemically active metals of s-block elements and less active metals of groups 13 and 14 and thus take their familiar name "transition elements"

The *f*-Block Elements (Inner-Transition elements)

The two rows of elements at the bottom of the periodic table, called the **Lanthanoids**$_{58}$Ce-$_{71}$Lu and **Actinoids.**$_{90}$Th-$_{103}$Lr are characterized by the outer electronic configuration $(n-2) f^{1-14} (n-1) d^{0-10}ns^2$. The last electron added to each element is an f-electron. These two series of elements are hence called the inner transition elements (f-Block Elements). They are all metals within each series, the properties of the elements are quite similar. The chemistry of the early actinoids is more complicated than the corresponding lanthanoids, due to the large number of oxidation states possible for these actinoid elements. Actinoid elements are radioactive. Many of the actinoid elements have been made only in nanogram quantities or less by nuclear reactions and their chemistry is not fully studied. The elements coming after uranium are called **transuranium** elements.

Example 1

The elements Z=117 and 120 have not yet been discovered. In which family / group would you place these elements and also give the electronic configuration in each case.

Solution

We see from the periodic table that element with Z=117, would belong to the halogen family (group 17) and the electronic configuration would be. $[Rn]4f^{14}5d^{10}7s^27p^5$. The element with Z=120, will be placed in group 2 (alkaline earth metals), and will have the electronic configuration $[Uuo]8s^2$.

In addition to displaying the classification of elements into s-, p-, d-, and f-blocks, the periodic table shows another broad classification of elements based on their properties. The elements can be divided into **Metals** and **Non-metals**. Metals comprise more than 75% of all knownelements and appear on the left side of the Periodic Table. Metals are usually solids at room temperature (Mercury is an exception); they have high melting and boiling points. They are good conductors of heat and electricity. They are malleable (can be flattened into thin sheets by hammering) and ductile (can be drawn into wires). In contrast non-metals are located at the top right hand side of the Periodic Table. Non-metals are usually solids or gases at room temperature with low melting and boiling points. They are poor conductors of heat and electricity. Most non-metallic solids are brittle and are neither malleable nor ductile. The elements become more metallic as we go down a group; the non-metallic character increases as one goes from left to right across the Periodic Table. The change form metallic to non-metallic character is not abrupt as shown by the thick zig-zag line in the periodic table. The elements (e.g. germanium, silicon, arsenic, antimony and tellurium) bordering this line and running diagonally across the Periodic Table show properties that are characteristic of both metals and non-metals. These elements are called **Semi Metals** or **Metalloids**.

Example 2

Arrange the following elements in the increasing order of metallic character: Si, Be, Mg, Na, P.

Solution

Metallic character increases down a group and decreases along a period as we move from left to right. Hence the order of increasing metallic character is, P<Si<Be<Mg<Na.

4.4 Periodicity of properties

Repetition of properties of elements at regular intervals is called periodicity in properties. The periodicity is due to similar electronic configuration of outer-most shells. Some of the properties are discussed below.

(i) Atomic and ionic radii

The size of an atom can be visualized from its atomic radius. The term atomic or ionic radius is generally defined as the distance between the centers of the nucleus and the outermost shell of electrons in an atom or ion. For example, the atomic radius of hydrogen atom is equal to 74/2 pm = 37 (bond distance in hydrogen molecule (H_2) is 74pm).

Atomic and ionic radii both decrease from left to right across a period in the periodic table when we consider only normal elements, e.g. in the elements of 2nd period the covalent radii decrease as we move from Li to F as shown below:

Elements of 2nd period:

Li Be B C N O F

Covalent radii ——— Values decreasing →

Thus in any period the alkali metals (that are present at the extreme left of the periodic table) have the largest size while the halogens (that are present at the extreme right, excluding the zero group elements) have the smallest size.

Explanations

We know that as we proceed from left to right in a period, the electrons are added to the orbitals of the same main energy level. Addition of different electrons to the same main energy level puts the electrons, on the average, no farther from the nucleus and hence the size cannot be increased. But with the addition of each electron, the nuclear charge (i.e. atomic number) increases by one. The increased nuclear charge attracts the electrons more strongly close to the nucleus and thus decreases the size of the atoms.

Table 4.5 Atomic Radii/pm Across the periods

Atom	Atomic radius	Atom	Atomic radius
Li	152	Na	186
Be	111	Mg	160
B	88	Al	143
C	77	Si	117
N	70	P	110
O	74	S	104
F	72	Cl	99

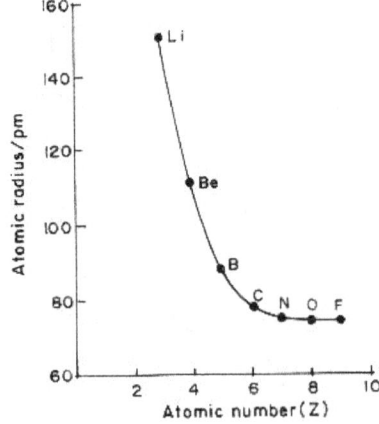

Fig. 4.1 Variation of atomic radius with atomic number across the second period

(b) **In a group**

On moving down a group of regular elements both atomic and ionic radii increase with increasing atomic number, e.g. in the elements of IA Group both covalent and ionic radii of M^+ ions increase when we pass from Li to Cs

Elements of IA Group :   Li Na K Rb Cs
Covalent radii/Ionic radii   Values increasing

Explanation

On proceeding downwards in a group the electrons are added to higher main energy levels, which are, on the average, farther from the nucleus. This effect decreases the electrostatic attraction between the nucleus and the valence-shell electrons and this decreased electrostatic attraction increases the atomic and ionic radii.

Table 4.6 Atomic Radii/pm Down the Group Across a Family

Atom	Atomic Radius	Atom	Atomic radius
Li	152	F	72
Na	186	Cl	99
K	231	Br	114
Rb	244	I	133
Cs	262	At	140

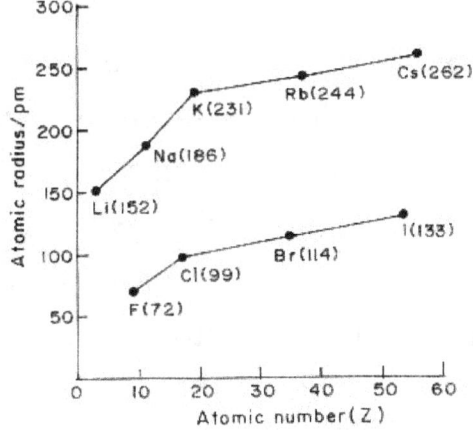

Fig. 4.2 Variation of atomic radius with atomic number for alkali metals and halogens

When we find some atoms and ions, which contain the same number of electrons, we call them **isoelectronic**. For example, O^{2-}, F^-, Na^+ and Mg^{2+} have the same number of electrons (10). Their radii would be different because of their different nuclear charges. The cation with the greater positive charge will have a smaller radius because of the greater attraction of the electrons to the nucleus. Anions with the greater negative charge will have the larger radius. In this case, the net repulsion of the electrons will outweigh the nuclear charge and the ion will expand in size.

Example

Which of the following species will have the largest and the smallest size Mg, Mg^{2+}, Al, Al^{3+}?.

Solution

Atomic radii decrease across a period. Cations are smaller than their parent atoms. Among isoelectronic ions, the one with the large positive nuclear charge will have a smaller radius.

Hence the largest species is Mg; the smallest one is Al^{3+}

The size of an anion greater while that of the cation is smaller than that of its parent atom, e.g. F^- (=1.36 Å)>F(=0.72 Å); Cl^-(=1.81 Å)>Cl(=0.99Å); Na^+(=0.95Å)<Na(=1.90Å); Ca^{2+}(=0.99 Å)<Ca(=1.97Å).

Explanation

Let us consider the radii of Na, Na^+, Cl and Cl^-. The reason of the fact that Na^+ ion is smaller than Na atom is that Na^+ ion has 10 electrons ($Na^+ \rightarrow 1s^2, 2s^2 p^6$) while Na atom has 11electrons (Na $\rightarrow 1s^2, 2s^2 p^6, 3s^1$). The nuclear charge (charge on the nucleus) in each case is the same, i.e. equal to +11 (atomic number of Na). This nuclear charge of +11 can pull 10 electrons of Na^+ ion inward more effectively than it can pull a greater number of 11 electrons of Na atom. Thus Na^+ ion is smaller than Na atom.

The reason why Cl^- ion is bigger than Cl atom can also be explained on a similar basis. The Cl^- ion has 18 electrons ($Cl^- \rightarrow 1s^2, 2s^2 p^6, 3s^2 p^6$) while Cl atom has only 17 electrons

$(Cl \rightarrow 1s^2, 2s^2p^6, 3s^2p^5)$. The nuclear charge in each case is +17, which cannot pull 18 electrons of Cl^- ion as effectively as it can pull 17 electrons of Cl atom inward. Thus Cl^- ion is bigger than Cl atom.

(ii) Ionization Energy: (Ionization Potential)

In modern terminology, ionization energy is known as **ionizationenthalpy**. The energy required to remove an electron from an atom isknown as ionization enthalpy (IE). The first ionization enthalpy may be defined as the amount of energy required to remove the most loosely bound electron from the isolated gaseous atom.

$$Atom_{(g)} + Energy \rightarrow Positive\ ion_{(g)} + Electron$$

For example,

$$Li_{(g)} + 520\ kJ\ mol^{-1} \rightarrow Li^+_{(g)} + e^-$$

Ionization enthalpy is also called *ionization potential* because it is measured as the amount of potential required to remove the most loosely held electron from the gaseous atom. It is expressed in terms of either kJ/mol or electron Volts/atom.

If a second electron is to be removed from the same element the energy required will be higher than that required for removal of the first electron because it is more difficult to remove an electron from a positively charged species than from a neutral atom.

$$Li^+_{(g)} + 7297\ kJ\ mol^{-1} \rightarrow Li^{2+} + e^-$$

Similarly the third ionization enthalpy will be higher than the second and so on. Fig. 4.3 shows a plot of first ionization enthalpy of some elements.

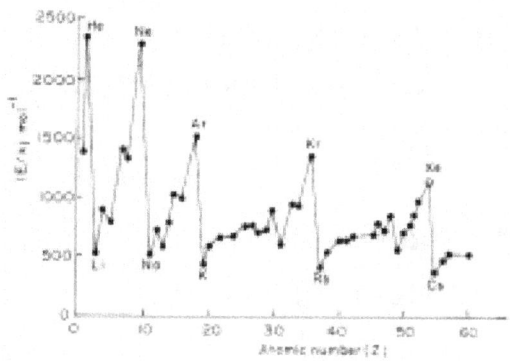

Fig. 4.3 Variation of first ionization analysis with atomic number for elements with Z = 1 to 60

Variation of Ionization Energy in the periodic Table

It is seen from the Fig. 4.4 that the ionization enthalpy of an element depends on its electronic configuration. Ionization potentials of noble gases are found to be maximum and those of alkali metals are found to minimum. The high values of noble gases are due to completely filled electronic configurations in their outermost shells and the low values of alkali metals are due to their large size and a single electron in the outermost shell.

The ionization potentials increases from left to right in a period. This trend can be explained in terms of increase in nuclear charge and decrease in size from left to right in a period. Generally the first ionization enthalpy decreases down a group in the periodic table. As we move down the group, the outer electrons, which are to be removed, are farther from the nucleus and there is an increasing screening of the nuclear charge by the electrons in the inner shells. Consequently the removal of electrons becomes easier down the group.

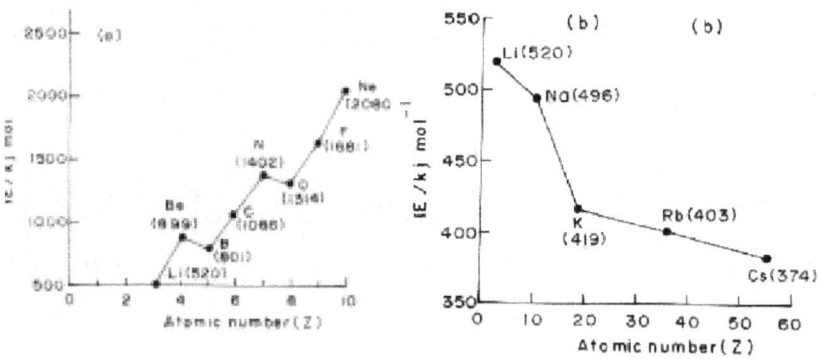

Fig. 4.4 (a) First ionization enthalpies of elements of the second period as a function of atomic number (b) First ionization enthalpies of alkali metals as a function of atomic number

Factors Influencing Ionization Enthalpy

The ionization enthalpy of an atom depends on the following factors.

(i) Size of the atom

As the distance between the electron and the nucleus increases, i.e., as the size of the atom increases, the outermost electrons are less tightly held by the nucleus. Thus, it becomes easier to remove an outermost electron. Thus ionization enthalpy decreases with increases in atomic size.

(ii) Charge on the nucleus

Ionization enthalpy increases with increase in nuclear charge because of the increase in the attractive force between the nucleus and the electron.

(iii) Screening effect of inner electrons

Ionization enthalpy decreases when the shielding effect of inner electrons increases. This is because when the inner electron shells increases, the attraction between the nucleus and the outermost electron decreases.

(iv) Penetration effect of electrons

The penetration power of the electrons in various orbitals decreases in a given shell (same value of n) in the order: s>p>d>f. Since the penetration power of s-electron towards the nucleus is more, it will be closer to the nucleus and will be held firmly. Thus, for the same shell, the ionization enthalpy would be more to remove the s-electrons in comparison with the p-electron which in turn would be more than that for d-electron and so on.

(v) Effect of half-filled and completely filled sub-levels

If an atom has half-filled or completely filled sub-levels, its ionization enthalpy is higher than that expected normally from its position in the periodic table. This is because such atom, have extra stability and hence it is difficult to remove electrons from these stable configurations.

(iii) Electron affinity

In modern terminology, electron affinity is known as the electron gain enthalpy. Electron gain enthalpy is the amount of energy released when an isolated gaseous atom accepts an electron to form a monovalent gaseous anion.

$$Atom_{(g)} + Electron \rightarrow Anion_{(g)} + energy$$

Example, $Cl_{(g)} + e^- \rightarrow Cl^-_{(g)} + EA$

If an atom has high tendency to accept an electron, large energy will be released. Consequently, electron gain enthalpy will be high. On the other hand if an atom has less tendency to accept the electron small amount of energy will be released, leading to small value of electron gain enthalpy. The values of electron gain enthalpy are expressed either in electron volt per atom or kilo joules per mole of atoms. For example, electron gain enthalpy of

F $= 322$ kJ mol^{-1}

Cl $= 349$ kJ mol^{-1}

   $= 324$ kJ mol^{-1}

Br and

I $= 295$ kJ mol^{-1}

Halogens (elements of group 17) can take up an electron to acquire the stable noble gas configuration. Their values for electron gain enthalpy are thus very high. Electron gain enthalpy values for the halogens are as in Fig. 4.5.

Electron gain enthalpies generally decrease on moving down the group. This is expected on account of the increase in size of atoms, the effective nuclear attraction for electrons decreases. As a result, there is less tendency to attract additional electrons with an increase in atomic number down the group.

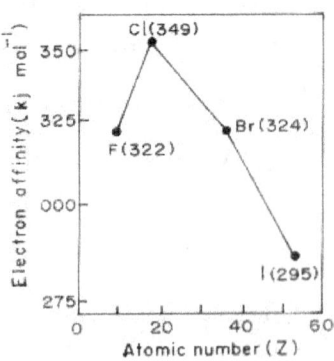

Fig. 4.5 Electron affinity enthalpies of halogens

From the electron gain enthalpy data of halogens it is clear that, contrary to expectation, the electron gain enthalpy of fluorine is lower than that of chlorine. This is because the fluorine atom has a very compact electronic shell due to its small size. The compactness of the fluorine shell results in electron repulsion whenever an electron is introduced into its 2p-orbital. This is why its electron gain enthalpy is less than the expected value. In Cl atom, 3p-orbitals are not as compact as the 2p-orbitals in fluorine atom. The incoming electron is more readily accepted by the chlorine atom because of weaker electron-electron repulsion. The electron gain enthalpy of chlorine is, therefore, higher than that of fluorine.

In the case of noble gases, the outer s-and p-orbitals are completely filled. No more electrons can be accommodated in these orbitals. *Noblegases, therefore, show no tendency to accept electrons.* Their electrongain enthalpies are zero.

Electron gain enthalpies generally increase as we move across a period from left to right. This is due to the increase in the nuclear charge, which results in greater attraction for electrons.

The second electron gain enthalpy refers to a process in which the electron is added to a negative ion. For example:

$$O^-_{(g)} + e^- \rightarrow O^{2-}_{(g)}$$

Since a negative ion O^- and an electron repel each other, energy is required and not released by the process. Therefore the second electron gain enthalpy is negative in this case.

Factors influencing the magnitude of electron affinity

The magnitude of EA is influenced by a number of factors such as(i) *Atomic size*; (ii) *Effective nuclear charge*; and (iii) *Screening effect byinner electrons*.

Example

Which of the following will have the most negative electron gain enthalpy and which has the least negative? P, S, Cl, F.

Explain your answer.

Solution

Electron gain enthalpy generally becomes more negative across a period as we move from left to right. Within a group, electron gain enthalpy becomes less negative down a group. However, adding an electron to the 2p orbital leads to greater repulsion than adding an electron to the larger 3p orbital. Hence the element with most negative electron gain enthalpy is chlorine; the one with the least negative electron gain enthalpy is phosphorus.

(iv) Electronegativity

Electronegativity may be defined as the tendency of an atom in a molecule to attract towards itself the shared pair of electrons. The mainfactors, which the electronegativity depends, are effective nuclear charge and atomic radius. Greater the effective nuclear charge greater is the electronegativity. Smaller the atomic radius greater is the electronegativity.

In a period electronegativity increases in moving from left to right. This is due to the reason that the nuclear charge increases whereas atomic radius decreases as we move from left to right in a period. Halogens have the highest value of electronegativity in their respective periods.

In a group electronegativity decreases on moving down the group. This is due to the effect of the increased atomic radius. Among halogens fluorine has the highest electronegativity. In fact fluorine is the most electronegative element and is given a value of 4.0 (Pauling's scale) whereas cesium is the least electronegative element (E.N. = 0.7) because of its largest size and maximum screening effect. In other words, cesium is the most electropositive element and hence is the most metallic element in the periodic table.

The main differences between Electron gain enthalpy (electro affinity) and electronegativity are given below:

Electron gain Enthalpy	Electronegativity
It is the tendency of an Isolated gaseous atom to attract an electron.	It is the tendency of an atom in a molecule to attract the shared pair of electrons.
It is measured in electron volts/atom or kcal/mole or	It is a number and has no units.

kj/mole. It is the property of an isolated atom. An atom has an absolute value of electron gain enthalpy. It does not change regularly in a period or group.	It is property of a bonded atom. An atom has a relative value of electronegativity depending upon its bonding state. For example, sp-hybridized carbon is more electronegative than sp^2-hybridized carbon which, in turn, is more electronegative than sp3-hybridized carbon. It changes regularly in a period or a group.

Electronegativity of an element is important in determining the bond character. If two atoms have the same electronegativity the bond between the two will be covalent, while a large difference in electronegativity leads to ionic bond. Between the extremes the purely covalent bond and purely ionic, the bonds will have different degrees of ionic character. As a rough estimate it is seen that a difference of 1.7 in electronegativity, the bond has 50% ionic character. If the difference is less than 1.7, the bond is considered covalent, and greater than 1.7 it is considered ionic.

4.5 Anomalous periodic properties in terms of screening constant, stability etc.

According to Hund's rule atoms having half-filled or completely filled orbitals are comparatively more stable and hence more energy is needed to remove an electron from such atoms. The ionization potentials of such atoms are, therefore, relatively higher than expected normally from their position in the periodic table.

Example

A few irregularities that are seen in the increasing values of ionization potential along a period can be explained on the basis of the concept of half-filled and completely filled orbitals, e.g., Be and N in the second period and Mg and P in the third period have slightly higher values of ionization potentials than those normally expected. This is explained on the basis of extra stability of the completely-filled 2s-orbital in Be(Be→$2s^2$) and 3s-orbital in Mg (Mg→$3s^2$) and of half-filled 2p-orbital in N (N→$2s^2p^6$) and 3p-orbital in P (P→$3s^2p^3$).

Another example for irregularity in Ionization potential is observed in the case of B and Be.

Ionization energy of boron (B →$2s^22p^1$) is lower than that of beryllium (Be → $2s^2$) [B = 8.3 eV, Be = 9.3 eV], since in case of boron we have to remove a $2p^1$ electron to get B^+[B ($2s^2p^1$) → $B^+(2S^2) + e^-$] while in case of Be we have to remove a $2s^1$ electron of the same main energy level to have Be^+ ion. [Be ($2s^2$) → $Be^+(2s^1)$ +e$^-$].

There is an exception to the vertical trend of ionization potential. This exception occurs in the case of those elements whose atomic numbers are greater than 72. Thus the ionization potentials of the elements from Ta_{73} to Pb_{82} are greater than those of the elements of the same sub-group above them as shown below: (First ionization potential values are given in electron volts, eV).

Table 4.8

V B	VI B	VII B	VIII			I B	II B	III A	IV A
Nb_{41}	Mo_{42}	Tc_{43}	Ru_{44}	Rh_{45}	Pd_{46}	Ag_{47}	Cd_{48}	In_{49}	Sn_{50}
6.8	7.1	7.2	7.3	7.4	8.3	7.5	8.9	5.7	7.3
Ta_{73}	W_{74}	Re_{75}	Os_{76}	Ir_{77}	Pt_{78}	Au_{79}	Hg_{80}	Tl_{81}	Pb_{82}
7.7	7.8	7.8	8.7	9.2	9.0	9.2	10.4	6.1	7.4

The reason for the abnormal behaviour (i.e. an increase in the value of I_1 from Nb → Ta, Mo → W, ...,Sn → Pb) shown by the elements from Ta_{73} to Pb_{82} is due to the lanthanide contraction as a result of which there occurs an increase in the nuclear charge without a corresponding increase in size through the rare earths. In fact, the size actually decreases in this region.

Periodic Variations

Similarly in moving down a group electron affinity values

generally decrease, e.g. $E_{Cl} > E_{Br} > E_I$. This is due to the steady increase in the atomic radius of the elements.

Exceptions

There are, however, some exceptions to this general rule as is evident from the following examples:

It is known that $E_F < E_{Cl}$ (E_F = 322 kJ mol^{-1}, E_{Cl} = 349 kJ mol^{-1}). The lower value of E for F is probably due to the electron-electron repulsion in relatively compact $2p$-orbital of F-atom.

In period, electron affinity values generally increase on moving fromleft to right in a period in the periodic table.

Exceptions

There are, however, exceptions also to this general rule; e.g.

Be and Mg have their E_A values equal to zero. Since Be and Mg have completely filled s-orbitals (Be $\rightarrow$ $2s^2$, Mg $\rightarrow$ $3s^2$), the additional electron will be entering the $2p$-orbital in case of Be and $3p$-orbital in case of Mg which are of considerably higher energy than the $2s$-and 3s orbitals respectively.

SUMMARY

In this Chapter, Historical Survey and review of Periodic Classifications are presented. Starting from Dobereiner law of triads, Newlands law of octaves, Mendeleev's periodic law and table are reviewed. Modern Periodic table is explained in detail. Correlating electronic configuration of elements justifies their positions in the modern periodic table. Latest IUPAC nomenclature of elements is also explained for elements having atomic number greater than 100 is explained.

Periodicity of properties likes Atomic and ionic radii, ionization energy, electron affinity, and electronegativity is discussed. Their trends along the period and down the group are explained with suitable examples. Anomalous periodic properties in terms of screening effect and stability are discussed.

REFERENCE:

Concise. Inorganic Chemistry J.D.Lee, 3rd Edition, 1977.

CHAPTER – 5

---

s - BLOCK ELEMENTS

1. 1 s Block Elements
2. 2 s Block Elements

1. 1 s Block Elements
OBJECTIVES

After Studying this Chapter you will able to:

- *Know about the occurrence and isotopes of hydrogen.*
- *Understand the nature and application of different isotopes of hydrogen.*
- *Differentiate ortho and para forms of hydrogen.*
- *Gain knowledge about the application of heavy water.*
- *Know about the preparation, properties and uses of hydrogen peroxide.*
- *Understand the special feature of liquid hydrogen used as a fuel - hydrogen economy.*
- *Analyse general properties of alkali metals.*
- *Know about the basic nature of oxides and hydroxides.*
- *Learn the extraction of alkali metal - lithium and sodium.*
- *Recognise the properties and understands the uses of lithium and sodium.*

Hydrogen is the first element in the periodic table. It has the simplest electronic configuration $1s^1$. It contains one proton in the nucleus and one electron.

Isotopes:- Atoms of the same element having same atomic number but different mass number are called isotopes.

There are three isotopes for hydrogen with mass numbers 1, 2 and 3, each possessing an atomic number of one.

Table 5.1 Isotopes of hydrogen

S. No	Name	Symbol	Atomic number	Mass number	Number of Protons	Number of Neutrons	Percentage abundance
1.	Protium or hydrogen	$_1H^1$	1	1	1	0	99.984
2.	Deuterium or heavy hydrogen	$_1H^2$	1	2	1	1	0.016
3.	Tritium	$_1H^3$	1	3	1	2	10^{-15}

The structure of the three isotopes of hydrogen are

1. **Protium or ordinary hydrogen:** It is the common form of hydrogen. Itconsists of one proton in its nucleus and one electron revolving around it. It constitutes 99.984% of total hydrogen available in nature. Its mass number is one.

2. **Deuterium or heavy hydrogen:** $1H^2 or 1D^2$. It occurs naturally in verysmall traces. The proportion present in naturally occurring hydrogen is in the approximate ratio: D: H~ 1:6000. It's nucleus consists of a proton and a neutron. However only a solitary electron is revolving around the nucleus. Its chemical properties are similar to those of protium but their reaction rates are different.

3. **Tritium, $_1H^3$ or $_1T^3$:** It occurs in the upper atmosphere only whereit is continuously formed by nuclear reactions induced by cosmic rays. Unlike deuterium, it is radioactive, with a half-life of ~ 12.3 years. It's nucleus consists of one proton and two neutrons.

They will have same similar chemical properties, however, their reaction rates will be different and their physical properties differ appreciably.

Methods of Preparation of deuterium

1. By Diffusion Process: It is possible to obtain deuterium directlyfrom hydrogen gas by taking advantage of different rates of diffusion of the two isotopes. The lighter hydrogen diffuses more quickly than deuterium through a porous partition under reduced pressure. Lower the pressure, higher is the efficiency of the process.

The process of diffusion has been carried out in various diffusion chambers called Hertz diffusion units. Each diffusion units consists of a porous membrane.

When the mixture is led into the diffusion units under reduced pressureby the help of mercury diffusion pumps, the heavier deuterium diffuses less readily while lighter hydrogen diffuses at faster rates. This process is repeated several times till hydrogen gets collected on the left while deuterium on the right. The efficiency of this process could be increased by increasing the number of diffusion units.

2.   By fractional distillation of liquid hydrogen: By fractionaldistillation of liquid hydrogen, it is possible to result in enrichment of the last fraction in deuterium because deuterium boils at 23.5K while hydrogen boils at lower temperature of 20.2K.

3.   By electrolysis of heavy water: As water contains about one partof heavy water in 6000 parts, at first, the concentration of heavy water is increased by fractional electrolysis of water containing an alkali between nickel electrodes. For example 1 ml of heavy water is obtained from about 20 litres by this method.

From heavy water, it is possible to get deuterium by decomposing it with sodium, red hot iron or tungsten or by its electrolysis containing sodium carbonate.

$$2D_2O \xrightarrow{\text{Electrolysis}} 2D_2 + O_2$$
$$\text{Deuterium}$$
$$D_2O + 2Na \longrightarrow NaOD + D_2$$
$$\text{Sodium deuteroxide.}$$

Deuterium obtained can be further obtained in purified state by the diffusion process.

Physical properties

Like hydrogen, deuterium is a colourless, odourless and tasteless gas which is insoluble in water and bad conductor of heat and electricity. The values of boiling point, melting point, vapour pressure, dissociation energy and latent heat of fusion are found to be lower for protium than deuterium.

Chemical properties

By virtue of its larger mass, deuterium reacts slower than hydrogen.

1. Burning in oxygen: Like hydrogen, it is combustible and burns inoxygen or air to give deuterium oxide which is also known as heavy water.

$$2D_2 + O_2 \longrightarrow 2D_2O.$$

2. Reaction with halogens: Like hydrogen, it combines with halogensunder suitable conditions to form their deuterides.

$$D_2 + Cl_2 \xrightarrow{\text{in light}} 2DCl$$
$$\text{Deuterium chloride}$$

$$D_2 + F_2 \xrightarrow{\text{in dark}} D_2F_2$$
$$\text{Deuterium fluoride}$$

3. Reaction with nitrogen: Like hydrogen, it combines with nitrogenin the presence of a catalyst to form nitrogen deuteride which are also known as heavy ammonia or deutero ammonia.

$$3D_2 + N_2 \longrightarrow 2ND_3$$

4. Reaction with metals: Like hydrogen, it reacts with alkali metals athigh temperatures (633K) to form deuterides

$$2Na + D_2 \longrightarrow 2NaD$$

5. Addition reactions: Like hydrogen, it gives addition reactions withunsaturated compounds. For example, a mixture of deuterium and ethylene when passed over heated nickel, gives Ethylene deuteride which is saturated hydrocarbon like ethane.

$$C_2H_4 + D_2 \xrightarrow[\text{535 K}]{\text{Ni}} CH_2D\text{-}CH_2D$$

6. Exchange reactions: Deuterium and hydrogen atoms undergoready exchange with H_2, NH_3, H_2O and CH_4 deuterium slowly exchanges their hydrogens partially or completely at high temperatures.

$$H_2 + D_2 \rightleftharpoons 2 HD$$
$$2NH_3 + 3D_2 \rightleftharpoons 2ND_3 + 3H_2$$
$$H_2O + D_2 \rightleftharpoons D_2O + H_2$$
$$CH_4 + 2D_2 \rightleftharpoons CD_4 + 2H_2$$
$$C_2H_6 + 3D_2 \rightleftharpoons C_2D_6 + 3H_2$$

Uses of deuterium

1. It is used as tracers in the study of mechanism of chemical reactions.
2. High speed deuterons are used in artificial radioactivity.
3. Its oxide known as heavy water (D_2O) which is employed as moderator in nuclear reactor to slow down the speed of fast moving neutrons.

Tritium $1H^3$: It is a rare isotope of hydrogen. Its traces are found innature due to nuclear reactions induced by cosmic rays.

Tritium is prepared by

i)   By bombarding lithium with slow neutrons

$$_3Li^6 + _0n^1 \longrightarrow {}_1T^3 + _2He^4$$

ii) By bombarding beryllium with deuterons

$$_4Be^9 + _1D^2 \longrightarrow {}_1T^3 + _4Be^8$$
$$_4Be^9 + _1D^2 \longrightarrow {}_1T^3 + 2\,_2He^4$$

Properties: It is radioactive with a half-life of 12.4 years. It decaysinto helium -3 with the emission of beta radiation.

$$_1T^3 \longrightarrow {}_2He^3 + _{-1}e^0$$

Uses

(i)   It is used as a radioactive tracer in chemical research.

(ii) It is used in nuclear fusion reactions.

5.1   Ortho and Para hydrogen

The nucleus of the hydrogen atom spins about an axis like a top. When two hydrogen atoms combine, they form molecular hydrogen. Thus depending on the direction of the two protons in the nucleus the following two types of hydrogen molecules are known. Hydrogen molecule in which both the protons in the nuclei of both H-atoms are known to spin in same direction is termed as ortho hydrogen. If the protons in the nuclei of both H-atoms spin in opposite direction, it is termed as parahydrogen.

At room temperature ordinary hydrogen consists of about 75% ortho and 25% para form. As the temperature is lowered, the equilibrium shifts in favour of para hydrogen. At 25K. There is 99% para and 1% ortho hydrogen. The change in the proportion of the two forms of hydrogen requires a catalyst such as platinum or atomic hydrogen or silent electric discharge.

The para form was originally prepared by absorbing ordinary hydrogen in activated charcoal in a quartz vessel kept at a temperature of 20K. The charcoal absorbs almost pure para hydrogen. By this method, pure para hydrogen can be isolated.

Conversion of para into ortho hydrogen

Ortho hydrogen is more stable than para hydrogen. The para form is transformed into ortho form by the following methods.

i.   By treatment with catalysts like platinum or iron
ii.   By passing an electric discharge
iii. By heating to 800°C or more.
iv.   By mixing with paramagnetic molecules like O_2, NO, NO_2.
v.   By mixing with nascent hydrogen or atomic hydrogen.

Properties: Ortho and para hydrogen are similar in chemical propertiesbut differ in some of the physical properties.

(i) Melting point of para hydrogen is 13.83K while that of ordinary hydrogen is 13.95 K.

(ii) Boiling point of para hydrogen 20.26K while that of ordinary hydrogen is 20.39K.

(iii) The vapour pressure of liquid para hydrogen is higher than that of ordinary liquid hydrogen.

(iv) The magnetic moment of para hydrogen is zero since the spins neutralize each other while in the case of ortho, it is twice than that of a proton.

(v) Para hydrogen possesses a lower internal molecular energy than ortho form.

5.2   Heavy water

It is also called as deuterium oxide. The oxide of heavy hydrogen (deuterium) is called heavy water. Heavy water was discovered by Urey in 1932. By experimental data he showed that `ordinary water', H_2O contains small proportion of heavy water, D_2O (about 1 part in 5000).

Preparation: The main source of heavy water is the ordinary waterfrom which it is isolated. Generally it is prepared by exhaustive electrolysis.

Principle: The heavy water is isolated either by prolonged electrolysisor by fractional distillation of water containing alkali. Taylor, Eyring and First in 1933 formulated the electrolysis of water in seven stages using N/2-NaOH solution and strip nickel electrodes.

The cell consists of a steel cell 18 inches long and 4 inches in diameter. The cell itself serves as the cathode while the anode consists of a cylindrical sheet of nickel with a number of holes punched in it. A large number of such cells are used for electrolysis of water in several stages. The gases obtained from each stage are separately burnt and the water thus formed is returned to the previous stage. The heavy water gradually concentrates in the residue left behind. The process usually consists of five stages.

A partial separation of heavy water from ordinary water can be affected by fractional distillation. This method utilizes the small difference in boiling points of protium oxide (H_2O) and deuterium oxide (D_2O).

Comparison of water and heavy water

Property	H₂O	D₂O
Density at $20^{\circ}C$	0.998	1.017
Freezing point	$0^{\circ}C$	$3.82^{\circ}C$
Boiling point	$100^{\circ}C$	$101.42^{\circ}C$
Maximum density	1.000 ($4^{\circ}C$)	1.1073 ($11.6^{\circ}C$)
Specific heat at $20^{\circ}C$	1.00	1.01
Surface tension at $20^{\circ}C$	72.8 dynes/cm	67.8 dynes/cm
Dielectric constant	82.0	80.5
Viscosity at $20^{\circ}C$	10.09 millipoises	12.6 millipoises

The solubilities of substances in heavy water also differ from those in ordinary water. Thus sodium chloride is about 15% less soluble in heavy water than in ordinary water.

Physical Properties

Heavy water is a colourless, odourless and tasteless mobile liquid. Higher viscosity of heavy water is responsible for lower solubility of ionic solids like NaCl and smaller mobility of ions.

Chemical Properties

The difference in chemical behaviour between H_2O and D_2O is very slight. However, the reaction velocity in general is slightly less in case of D_2O reactions.

Important reactions of heavy water

1. With metals

    D_2O reacts slowly with alkali and alkaline earth metals liberating heavy hydrogen.

    $2Na + 2D_2O \longrightarrow 2NaOD + D_2$
    Sodium deuteroxide

    $Ca + 2D_2O \longrightarrow Ca(OD)_2 + D_2$
    Calcium deuteroxide

2. With metallic oxides

    Metals like sodium and calcium dissolve in D_2O and form heavy alkalies.

    $Na_2O + D_2O \rightarrow 2NaOD$

    $CaO + D_2O \rightarrow Ca(OD)_2$

3. With acid anhydrides

    D_2O forms corresponding acids containing heavy hydrogen.

    $SO_3 + D_2O \rightarrow D_2 + SO_4$
    Deutero sulphuric acid

    $P_2O_5 + 3D_2O \rightarrow 2D_3PO_4$
    Deuterophosphoric acid

4. Upon electrolysis, heavy water containing dissolved P_2O_5, decomposes into deuterium and oxygen which are liberated at the cathode and anode respectively.

$$2D_2O \longrightarrow 2D_2 + O_2$$

5. With salt and other compounds they form deuterates.

$$CuSO_4.5D_2O, \ Na_2SO_4.10D_2O, \ NiCl_2.6D_2O$$

7. Exchange reactions

When compounds containing hydrogen are treated with D_2O, hydrogen undergoes an exchange for deuterium.

$$NaOH + D_2O \longrightarrow NaOD + HOD$$
$$NH_4Cl + 4D_2O \longrightarrow ND_4Cl + 4HOD$$

Biological Properties

In general heavy water, retards the growth of living organisms like plants and animals. The tobacco seeds do not grow in heavy water. Also, pure heavy water kills small fish, tadpoles and mice when fed upon it. Certain moulds have been found to develop better in heavy water.

Uses of heavy water

1. As a neutron moderator, in nuclear reactors.
2. It is used as a tracer compound in the study of reactions occurring in living organisms.
3. It is used for the preparation of deuterium.

5.3   Hydrogen peroxide

Hydrogen peroxide was first prepared by L.J.Thenard, in 1813 by theaction of dilute acid on barium peroxide. Traces of H_2O_2 are found in atmosphere and in certain plants.

Laboratory preparation of hydrogen peroxide

1. By the action of dilute sulphuric acid on sodium peroxide. Calculated quantity of Na_2O_2 is added in small proportions to a 20% ice cold solution of sulphuric acid.

$$Na_2O_2 + H_2SO_4 \longrightarrow Na_2SO_4 + H_2O_2$$

2. Pure H_2O_2 is obtained by reacting BaO_2 with an acid

$$BaO_2 + H_2SO_4 \longrightarrow BaSO_4 + H_2O_2$$
$$3BaO_2 + 2H_3PO_4 \longrightarrow Ba_3(PO_4)_2 + 3H_2O_2$$

3. H_2O_2 is manufactured by followed by vacuum distillation. H_2O_2.

The electrolysis of 50% sulphuric acid The distillate is 30% solution of pure

Reactions

$$H_2SO_4 \rightleftharpoons H^+ + HSO_4^-$$

$$2HSO_4^- \rightleftharpoons H_2S_2O_8 + 2e^- \text{ (At anode)}$$

$$H_2S_2O_8 + H_2O \rightleftharpoons H_2SO_4 + H_2SO_5$$

$$H_2SO_5 + H_2O \rightleftharpoons H_2SO_4 + H_2O_2$$

$$2H^+ + 2e^- \rightleftharpoons H_2 \text{ (At cathode)}$$

Concentration of hydrogen peroxide solution

The impurities like organic material or metallic ions, may catalyze its explosive decomposition.

i) By careful evaporation of the solution obtained above on a water bath preferably under reduced pressure using fractionating column.

ii) By distillation under reduced pressure at temperatures below 330K, the concentration up to 90% solution is used till crystallization formed.

Strength of Hydrogen peroxide

The strength of a sample of hydrogen peroxide solution is expressed in terms of the volumes of oxygen at S.T.P that one volume of H_2O_2 gives on heating.

Properties

Physical

H_2O_2 is a colourless, odourless, syrupy liquid in the anhydrous state. It is miscible with water, alcohol, and ether in all proportions.

Chemical

Pure H_2O_2 is unstable and decomposes on standing. On heating when water and oxygen are formed.

$$2H_2O_2 \rightarrow 2H_2O + O_2$$

Oxidizing Properties

H_2O_2 is a powerful oxidizing agent. It functions as an electron acceptor.

$$H_2O_2 + 2H^+ + 2e^- \rightarrow 2H_2O$$

(In acidic solution)

$$H_2O_2^- + 2e^- \rightarrow 2OH^-$$
(In alkaline solution)

i) It oxidizes PbS to PbSO$_4$

$$PbS + 4H_2O_2 \rightarrow PbSO_4 + 4H_2O$$

ii) It oxidizes ferrous salts into ferric salts.

$$2Fe^{2+} + 2H^+ + H_2O_2 \rightarrow 2Fe^{3+} + 2H_2O$$

Due to its oxidizing property, it is a valuable bleaching agent, powerful but harmless disinfectant and germicide. Delicate materials like silk, wool, hair which will be destroyed by chlorine, are bleached with H$_2$O$_2$.

Reducing Properties

With powerful oxidizing agents, H$_2$O$_2$ acts as a reducing agent. Moist silver oxide, acidified KMnO$_4$, ozone, chlorine and alkaline solutions of ferricyanides are reduced.

$$Ag_2O + H_2O_2 \rightarrow 2Ag + H_2O + O_2$$

Uses

i) It destroys bacteria and hence it is used as an antiseptic and germicide for washing wounds, teeth and ears.

ii) It destroys the colour of some organic compounds and is used in bleaching delicate things like hair, wool, silk ivory and feathers.

iii) It is used as an oxidizing agent.

iv) It is also used as a propellant in rockets.

5.4 Liquid hydrogen as a fuel

The hydrogen atom has become a model for the structure of atom. Hydrogen as a substance however, has an equally important place in chemistry. Hydrogen is normally a colourless, odourless gas composed of H$_2$ molecules. Approximately 40% of the hydrogen produced commercially is used to manufacture ammonia and about the same amount is used in petroleum refining. But the future holds an even greater role for hydrogen as a fuel.

Liquid hydrogen, H$_2$, is a favourable rocket fuel. On burning, it produces more heat per gram than any other fuel. In its gaseous form, hydrogen may become the favourite fuel of the twenty first century.

When hydrogen burns in air, the product is simply water. Therefore, the burning of hydrogen rather than fossil fuels (natural gas, petroleum, and coal) has important advantages.

The burning of fossil fuels is a source of environmental pollutants. They become the source of acid rain and discharge a large amount of toxic gases like SO_2 and CO_2.

Controlling carbon dioxide emissions into the atmosphere is a difficultchallenge, but the answer might lie in the conversion to a hydrogen economy, hydrogen would become a major energy barrier. Automobiles, for example may be modified to burn hydrogen. At present, in USA they use car using a modified piston engine and has a hydrogen storage unit in the tank. This proves that it is possible to develop hydrogen-burning cars.

Hydrogen in not a primary energy source. But it is a convenient and non-polluting fuel, but it would have to be obtained from other energy sources.

1) It is produced by heating propane and steam at high temperature and pressure in presence of the catalyst nickel.

$$C_3H_{8(g)} + 3H_2O_{(g)} \xrightarrow{\text{Ni}} 3CO_{(g)} + 7H_{2(g)}$$

2) Pure hydrogen may be produced by reacting carbonmonoxide with steam in the presence of a catalyst to fix CO_2 and H_2. The CO_2 is removed by dissolving it in a basic aqueous solution.

3) Hydrogen can be obtained directly from water that is decomposed by some form of energy. For example, electricity from solar photovoltaic collectors can be used as a source of energy to decompose water by electrolysis. Researchers use solar energy to convert water directly to hydrogen and oxygen.

5.5 Alkali Metals

Position of alkali metals in the periodic table

Alkali metals occupy the group I of the periodic table. Elements lithium, sodium, potassium, rubidium, cesium and francium constitute alkali metals. They are named so from the Arabic word `Alquili' meaning `plant ashes'. Ashes of plants are composed mainly of sodium and potassium carbonates.

General characteristics

1. The alkali metals are shiny white and soft.
2. They can be readily cut with a knife.
3. They are extremely reactive metals and form strong alkaline oxides and hydroxides.
4. The last metal of this group, francium is radioactive.
5. Since the alkali metals are extremely reactive they occur only as compounds in nature.
6. All the alkali metals exhibit an oxidation state of +1. This is because the metals can easily lose their single outermost electron.
7. The alkali metals give characteristic colour in Bunsen flame. The colour given by Li, Na and K are crimson red, yellow, lilac respectively. This is because when the alkali metal or any of its compounds are heated in a Bunsen flame, the ns' electron gets excited to higher energy levels and while returning to their ground state the excitation energy absorbed by them is released as light in the visible region.

Table 5.2 Electronic configuration of alkalimetals

Element	Symbol	Atomic Number	Electronic configuration
Lithium	Li	3	[Helium] $2s^1$
Sodium	Na	11	[Neon] $3s^1$
Potassium	K	19	[Argon]$4s^1$
Rubidium	Rb	37	[Krypton]$5s^1$
Caesium	Cs	55	[Xenon]$6s^1$
Francium	Fr	87	[Radon]$7s^1$

Gradation in Physical Properties

1. **Density:** In general, these elements have high density due to theclose packing of atoms in their metallic crystals. Lithium has low density due to the low atomic weight of the atom. Density of the elements increases on moving down the group due to the increase in

the mass of the atoms with increasing atomic number. However, K is lighter than Na probably due to an unusual increase in atomic size.

2. **Atomic volume:** Atomic volume increases on moving down thegroup from Li to Cs. Hence there is an increase in atomic and ionic radii in the same order.

3. **Melting and boiling points:** All alkali metals have low melting andboiling point due to the weak bonding in the crystal lattice. The weak interatomic bonds are attributed to their large atonic radii and to the presence of one valence electron. With the increase in the size of the metal atoms, the repulsion of the non-bonding electron gets increased and therefore melting and boiling points decreases on moving down the group from Li to Cs.

4. **Ionization energy:** The first ionization energies of alkali metals arerelatively low and decreases on moving down from Li to Cs.

$$M_{(g)} \rightarrow M^+_{(g)} + 1e^-$$

As the atomic radius gets increased on moving down the group, the outer electron gets farther and farther away from the nucleus and therefore ionization energy decreases.

The second ionization energies of alkali metals are fairly high. This implies that the loss of the second electron is quite difficult, because it has to be pulled out from the noble gas core.

5. **Electropositive character:** As alkali metals have low ionizationenergies, they have a great tendency to lose electrons forming unipositive ions. Therefore they

$$M \rightarrow M^+ + 1e^-$$

have strong electropositive character. Electropositive character increases as we go down the group. The alkali metals are so highly electropositive that they emit electrons when irradiated with light. This effect is known as photoelectric effect. Due to this property, Cs and K are used in photoelectric cells.

6. **Oxidation state:** All the alkali metals have only one

electron intheir outermost valence shall. As the penultimate shell being complete, these elements lose one electron to get the stable configuration of the nearest inert gas. Thus, they are monovalent elements showing an oxidation state of +1.

7. **Reducing properties:** As alkali metals have low ionization energy,they lose their valence electrons readily and thus bring about reduction reaction. Therefore these elements behave as good reducing agents.

5.6 Extraction of Lithium and Sodium

Extraction of Lithium - Electrolysis of Lithium chloride.

Lithium metal is obtained by the electrolysis of moisture free lithium chloride in a crucible of thick porcelain using gas-coke splinter anode and iron wire cathode. For the preparation of the metal on a large scale, a fused mixture of equal parts of lithium and potassium chloride is used, as it melts at a lower temperature of 720 K.

Lithium is also obtained by the electrolysis of a concentrated solution of lithium chloride in pyridine or acetone.

Properties of Lithium

Physical: Lithium is a silvery white metal and it is the lightest of allsolid elements. It's vapours impart calamine red colour to the flame. It is a good conductor of heat and electricity. It gives alloys with number of metals and forms amalgam.

Chemical

1) With air: Lithium is not affected by dry air but in moist air it is readily oxidized. When heated in air above 450K, it burns to give lithium monoxide and lithium nitride

$$4\,Li + O_2 \longrightarrow 2\,Li_2O$$
$$6\,Li + N_2 \longrightarrow 2\,Li_3N$$

2) It decomposes cold water forming lithium hydroxide and hydrogen

$$2Li + 2H_2O \longrightarrow 2LiOH + H_2$$

3) Lithium is a strongly electropositive metal and displaces hydrogen from acid with the formation of corresponding lithium salts. Dilute

and concentrated hydrochloric and dilute sulphuric acid react readily while concentrated sulphuric acid reacts slowly. With nitric acid, the action is violent and metal melts and catches fire.

Uses

1) For the manufacture of alloys.
2) As a deoxidizer in the preparation of copper and nickel.
3) Lithium citrate and salicylate are used in the treatment of gout
4) $LiAlH_4$ is used as an reducing agent.
5) Its compounds are used in glass and pottery manufacture.

Extraction of Sodium

Down's process: It is now manufactured by electrolysis of fusedsodium chloride.

Down's electrolytic cell, consists of an iron box through the bottom of which rises a circular carbon anode. The anode is surrounded by a ring shaped iron cathode enclosed in a wire gauze shell which also acts as a partition and separates the two electrodes.

On electrolysis, chlorine is liberated at the anode and let out through an exit at the top. Sodium is liberated at the cathode and remains in the wire-gauze shell. Level of molten sodium rises and it overflows into a receiver.

$$2NaCl \longrightarrow 2Na + Cl_2$$

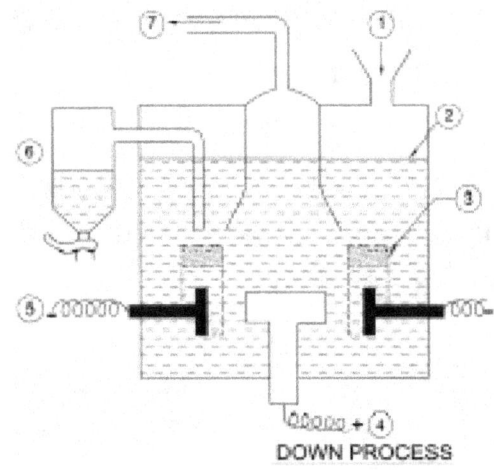

DOWN PROCESS

1	NaCl
2	FUSED NaCl
3	WIRE GAUZE SHELL
4	GRAPHITE ANODE
5	IRON CATHODE RING SHAPED
6	SODIUM
7	CHLORINE

Basics of Inorganic Chemistry

Physical properties

1) It is a silvery white metal when freshly cut but is rapidly tarnished in air. It forms tetragonal crystals.
2) It is a soft metal.
3) It is a good conductor of electricity.
4) It dissolves in liquid ammonia forming an intense blue solution.

Chemical

1) Action of air: In moist air a layer of sodium oxide, hydroxide andcarbonate is formed on its surface which loses its lustre.

$$4Na + O_2 \xrightarrow{} 2\,Na_2O \xrightarrow{2H_2O} 4NaOH \xrightarrow{2CO_2} Na_2CO_3 + 2H_2O$$

When heated in air, it burns violently to form the monoxide and the peroxide.

$$4Na + O_2 \xrightarrow{} 2Na_2O$$
$$2Na + O_2 \xrightarrow{} Na_2O_2$$

2) Action of water : It decomposes water vigorously, liberating hydrogen and forming sodium hydroxide.

$$2Na + 2H_2O \xrightarrow{} 2NaOH + H_2$$

3) Action of ammonia: Sodium gives sodamide with ammonialiberating hydrogen.

$$2Na + 2NH_3 \xrightarrow{570\text{-}670K} 2NaNH_2 + H_2.$$

Sodium dissolved in liquid ammonia is used as a reducing agent in organic chemistry.

4) Action of acids: It displaces hydrogen from acids

$$2HCl + 2Na \xrightarrow{} 2NaCl + H_2.$$

5) Reducing action: Reduces many compounds when heated withthem in the absence of air

$$Al_2O_3 + 6Na \xrightarrow{} 2Al + 3Na_2O$$
$$SiO_2 + 4Na \xrightarrow{} Si + 2Na_2O.$$

Reduces carbondioxide when heated forming carbon and sodium carbonate.

$$4Na + 3CO_2 \xrightarrow{} 2Na_2CO_3 + C.$$

6) With Mercury: When heated with mercury, sodium forms anamalgam of varying composition Na_2Hg, Na_3Hg, $NaHg$ etc.

Uses

1) For the preparation of sodium peroxide, sodamide and sodium cyanide, tetraethyl lead etc.
2) Sodium amalgam is employed as a reducing agent.
3) As a deoxidizing agent in the preparation of light alloys and some rare earth metals from their oxides.
4) It acts as a catalyst in the polymerization of isoprene (C_5H_3) into artificial rubber.
5) As a reagent in organic chemistry.

SUMMARY

Hydrogen is the first element in the periodic table. It exists in 3 isotopes. Protium, deuterium and tritium. The preparation properties of deuterium are dealt in detail.

Depending on the spins of the nucleus of hydrogen atom in a molecule, two types ortho and para hydrogen are known. It can be converted from one to another form.

One of the important compound of deuterium is heavy water, which is isolated from ordinary water. It reacts with metals, metallic oxides, acid anhydrides etc. It also undergo exchange reaction.

In 1813, L.J. Thenard prepared hydrogen peroxide by the action of dilute acid on barium peroxide. Traces of it are found in atmosphere. Pure H_2O_2 is unstable. It acts as a powerful oxidizing agent.

The use of liquid hydrogen as a fuel is explained in this chapter.

Group 1 elements are known as alkalimetals. There is mostly regular gradation in properties like density, atomic volume, melting and boiling point, ionization energy etc. along the group.

The extraction of lithium and sodium and its properties are explained in detail.

REFERENCES:

1. General Chemistry - John RussellMcGraw Hill International Editions 3rd Edition.
2. Inorganic Chemistry, P.L. Soni.

2.   2 s Block Elements
OBJECTIVES
After Studying this Chapter you will able to:

· *Understands the general characteristics of alkaline earth metals*
· *Know the comparison of alkali and alkaline earth metals.*
· *Understand the extraction of magnesium by electrolytic process, its properties and uses.*
· *Recognize the different compounds of alkaline earth metals.*
· *Understand and learn in detail about the preparation, properties and uses of CaO, plaster of paris and MgSO4.*

The second group of the periodic table contains Beryllium (Be),Magnesium (Mg), Calcium (Ca), Strontium (Sr), Barium (Ba) and Radium (Ra). These elements are also a known as "Alkaline Earth Metals". The word earth was applied in old days to a metallic oxide and because the oxides of calcium, strontium and barium produced alkaline solutions in water and, therefore these metals are called the alkaline earth metals. Radium corresponds to all the alkaline earth metals in its chemical properties but being radioactive, it is studied along with other radioactive elements.

Like the alkali metals, they are very reactive and hence never occur in nature in free form and react readily with many non metals.

Electronic configuration

Element	At No.	Electronic	Configuration of Valence Shell
Beryllium	4	$1s^2 2s^2$	$2s^2$
Magnesium	12	$1s^2 2s^2 2p^6 3s^2$	$3s^2$
Calcium	20	$1s^2 2s^2 2p^6 3s^2 3p^6 4s^2$	$4s^2$
Strontium	38	$1s^2 2s^2 2p^6 3s^2 3p^6 4s^2 4p^6 5s^2$	$5s^2$
Barium	56	$1s^2 2s^2 2p^6 3s^2 3p^6 3d^{10} 4s_2 4p^6 4d^{10} 5s^2 5p^6 6s^2$	$6s^2$
Radium	88	$1s^2 2s^2 2p^6 3s^2 3p^6 3d^{10} 4s^2 4p^6 4d^{10} 5s^2 5p^6 5d^{10} 5f^{14} 6s^2 6p^6 7s^2$	$7s^2$

The electronic configurations show that for each element, the neutral atom has two electrons after inert gas core and two electrons are in acompleted s-subshell. Thus, the outer electronic configuration of each element is ns^2 where n is the number of the valence shell. It can be expected that the two electrons can be easily removed to give the inert gas electronic configuration. Hence these elements are all bivalent and tend to form ionic salts. Thus ionic salts are less basic than group 1. Due to their alike electronic structure, these elements resemble closely in physical and chemical properties.

The variation in physical properties is not as regular as for the alkalimetals because the elements of this group do not crystallize with the same type of metallic lattice.

These elements have been sufficiently soft yet less than the alkalimetals as metallic bonding in these elements has been stronger than in first group alkali elements.

Beryllium is unfamiliar, partly because it is not very abundant and partly because it is difficult to extract. Magnesium and calcium are abundant and among the eight most common elements in the earth's crust. Strontium and barium are less abundant but are well known, while radium is extremely scarce and its radioactivity is more important than its chemistry.

Metallic properties

The alkaline earth metals are harder than the alkali metals. Hardness decreases, with increase in atomic number. They show good metallic luster and high electrical as well as thermal conductivity because the two s-electrons can easily move through the crystal lattice.

Melting and Boiling Points

Both melting and boiling points do not show regular trends because atoms adopt different crystal structures. They possess low melting and boiling points. These are, however, higher than those of alkali metals because the number of bonding electrons in these elements is twice as great as group 1 elements.

Atomic radius

The atoms of these elements are somewhat smaller than the atoms of the corresponding alkali metals in the same period. This is due to highernuclear charge of these atoms which tends to draw the orbital electrons inwards. Due to the smaller atomic radius, the elements, are harder, have higher melting points and higher densities than the elements of group 1. Atomic radius is seen to increase on moving down the group on account of the presence of an extra shell of electron at each step.

Ionic radius

The ions are also large but smaller than those of the elements in group 1. This is due to the fact that the removal of two orbital electrons in the formation of bivalent cations M^{2+}, (Be^{2+}, Mg^{2+}, Ca^{2+}, Sr^{2+}, etc) increases the effective nuclear charge which pulls the electrons inwards and thus reduces the size of the ions. The ionic radius is seen to increase on moving down the group 2.

Atomic volume

Due to the addition of an extra shell of electrons to each element from Be to Ra, the atomic volume increases from Be to Ra.

Ionization Energy

As the alkaline earth metals are having smaller size and greater nuclear charges than the alkali metals, the electrons are more tightly held and hence the first ionization energy would be greater than that of the alkali metal.

The second ionization energy has been to be nearly double than that of the first ionization energy.

It is interesting to observe that although the IE_2 of the alkaline earth metals is much higher than the IE_1 they are able to form, M^{2+}ions. This is due to their high heat of hydration in aqueous solution and high lattice energy in the solid state. As the atomic size gets increased from Be to Ba, the values of IE_1 and IE_2 of these elements would decrease on going down the group, i.e. Be to Ba.

As among second group elements beryllium has the highest ionization energy. It has the least tendency to form Be^{2+} ion.

Thus its compounds with nitrogen, oxygen, sulphur and halogens are covalent whereas the corresponding compounds of Mg, Ca, Sr and

Ba are ionic.

The total energy required to produce gaseous divalent ion for second group elements is over four times greater than the amount needed to ionize alkali metals. This very high energy requirement is more than offset by the hydration energy or the lattice energy being more than four times greater.

Oxidation states

Because of the presence of two s-electrons in the outermost orbital, being high heat of hydration of the dispositive ions and comparatively low value of IE_2, the alkaline earth metals have been bivalent. The divalent ion is having no unpaired electron, hence their compounds are diamagnetic and colourless, provided their anions have been also colourless.

Flame colouration

These elements and their compounds impart characteristic colour to flame. Thus, barium - apple green, calcium - brick red, strontium - crimson red, radium - crimson red.

The reason for imparting the colour to flame is that when elements or their compounds are put into flame, the electrons get energy and excite to higher energy levels. When they return to the ground state they emit the absorbed energy in the form of radiations having particular wavelength.

Beryllium and magnesium atoms are smaller and their electrons being strongly bound to the nucleus are not excited to higher energy levels. Therefore they do not give the flame test.

Diagonal relationship between Beryllium and Aluminium

In case of beryllium, a member of second period of the periodic table, which resembles more with Aluminium group (13 group) than the member of its own group (2nd). The anamolous behaviour of beryllium is mainly ascribed to its very small size and partly due to its high electronegativity. These two factors tend to increase the polarizing power of Be^{2+} tends to form ions to such extent that it is significantly equal to the polarizing power of Al^{3+} ions. Thus the two elements resemble very much.

5.7 Magnesium

The magnesium comes from the name of the mineral magnetite, which in turn is believed to stem from the name Magnesia. The British chemist Humphrey Davy discovered the pure element magnesium in 1808.

Due to its low density, it is considered to be a structural unit.

Important Ores

Magnesium does not occur in the native state. In the combined state it occurs very abundantly in the earth crust.

Magnesite, $MgCO_3$ Dolomite, $MgCO_3, CaCO_3$

Epsomsalt, $MgSO_4, 7H_2O$ Carnallite $MgCl_2 KCl.6H_2O$

However magnesium ion Mg^{2+}, is the third most abundant dissolved ion in the oceans, after Cl^- and Na^+. The oceans are the best sources for magnesium. It is widely distributed in the vegetable kingdom being present in chlorophyll, the green colouring matter of the leaves.

Metallurgy

Magnesium is prepared on a large scale by the electrolysis of either fused magnesium chloride or magnesia.

1. Electrolysis of fused magnesium chloride

The purified carnallite ore is the principal source for this process. A mixture of equal quantities of carnallite and NaCl is fused to a clear liquid at 973K. The alkali chloride prevents hydrolysis of magnesium chloride and increases the conductivity of the fused mass.

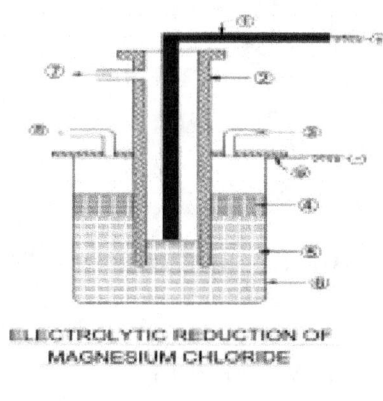

ELECTROLYTIC REDUCTION OF
MAGNESIUM CHLORIDE

1. GRAPHITE ANODE
2. PORCELAIN HOOD
3. INERT GAS LIKE COAL GAS
4. MAGNESIUM
5. ELECTROLYTE (MOLTEN)
6. IRON CELL
7. CHLORINE
8. INERT GAS
9. IRON CATHODE

The electrolysis of the fused mass is carried out in an atmosphere of coal gas in air tight iron cell which can hold 6-7 tones of the electrolyte. The temperature of the electrolyte bath is maintained at 970K. The iron cell itself acts as a cathode unlike the anode consists of a carbon or graphite rod surrounded by a porcelain tube through which the liberated chlorine escapes. Molten magnesium being lighter than the electrolyte rises to the surface and is periodically removed with perforated ladle. The electrolysis is carried out in an atmosphere of coal gas so as to avoid the oxidation of molten magnesium. The metal thus obtained is 99.9% pure. It may be further purified by remelting with a flux of anhydrous magnesium chloride and sodium chloride.

Physical

Pure magnesium metal is a relatively active silvery white metal. At slightly below its melting point, it is malleable and ductile and can be drawn into wire or rolled into ribbon in which form it is generally sold. It is a very light metal.

Chemical Properties

1. **Action of Air:** It does not tarnish in dry air but a layer of white oxide isformed on its surface in moist air.

2. **With air on burning:** It burns in air or oxygen with a dazzling lightrich in ultraviolet rays, forming magnesium oxide and magnesium nitride.

    $$2Mg + O_2 \rightarrow 2MgO$$
    $$3Mg + N_2 \rightarrow Mg_3N_2$$

3. **With CO_2**

    It continues to burn in CO_2,
    $$2\,Mg + CO_2 \rightarrow 2MgO + C$$

4. **Action of Water**

    When heated with steam it burns brilliantly producing magnesium oxide and hydrogen.
    $$Mg + H_2O \rightarrow MgO + H_2$$

steam

5. Action of Acids

Dilute HCl or H_2SO_4 gives hydrogen with magnesium. With dilute HNO3, part of the hydrogen liberated is oxidized by nitric acid, which itself is reduced to a variety of products depending upon the concentration. With concentrated HNO_3, it gives ammonium nitrate.

$$4Mg + 10HNO_3 \rightarrow 4Mg(NO_3)_2 + NH_4NO_3 + 3H_2O$$

6. Displacement of Metals

It is a strongly electropositive metal and hence Mg displaces nearly all the metals from the solutions of their salts eg.

$$Mg + 2AgNO_3 \rightarrow Mg(NO_3)_2 + 2Ag$$

7. Reducing Action

Mg has great affinity for oxygen and it liberates sodium, potassium, boron and silicon from their oxides at high temperatures.

$$K_2O + Mg \rightarrow MgO + 2K$$
$$B_2O_3 + 3Mg \rightarrow 3MgO + 2B$$

Uses of Magnesium

1. In flashlight photography, pyrotechnics and in fireworks.
2. As a reducing agent in the preparation of boron and silicon and deoxidizer in metallurgy.

5.8 Compounds of alkaline earth metals

Magnesium sulphate, epsom slat, $MgSO_4.7H_2O$

It is prepared by dissolving magnesium oxide or carbonate in dilute sulphuric acid.

$$MgO + H_2SO_4 \rightarrow MgSO_4 + H_2O$$

Uses

1) As a purgative
2) In dyeing and tanning processes and in dressing cotton goods.
   Platinized $MgSO_4$ is used as a catalyst.

Calcium oxide, CaO, quicklime

It is prepared by burning limestone in specially designed kilns.

$$CaCO_3 \xrightarrow{1070K} CaO + CO_2$$

Properties

1. Lime is a white porous solid
2. On adding water it gives a hissing sound and becomes very hot. The fine powder obtained is known as slaked lime and consists of calcium hydroxide $Ca(OH)_2$. This process is called slaking of lime.

$$CaO + H_2O \rightarrow Ca(OH)_2$$

The paste of lime in water is called milk of lime whereas the filtered and clear solution is known as lime water.

3. With chlorine it gives bleaching powder $CaOCl_2.H_2O$
4. With carbondioxide, it forms calcium carbonate while with sulphurdioxide, calcium sulphite is obtained.
5. Moist hydrochloric acid gas reacts with it to give calcium chloride but there is no action with the dry gas.

$$CaO + 2HCl \rightarrow CaCl_2 + H_2O$$

Uses

1. For the manufacture of calcium chloride, cement, mortar and glass.
2. For drying gases and alcohol.
3. As milk of lime, used in refining sugar and white washing.
4. As lime water, used as a reagent in laboratory and in medicine.

Calcium sulphate, $CaSO_4$

It occurs as Anhydrite, $CaSO_4$ and Gypsum $CaSO_4. 2H_2O$. It may be prepared by adding dilute sulphuric acid to the solution of a calcium salt.

$$CaCO_3 + H_2SO_4 \rightarrow CaSO_4 + H_2O + CO_2$$

Uses of Gypsum

It is used

1. For preparing plasters
2. As a retardant for the setting of cement

Plaster of Paris

When Gypsum is heated to about 393K it loses 1½ molecules of water and forms plaster of paris with the formula $CaSO_4 \cdot \frac{1}{2}H_2O$, Calcium Sulphate hemihydrate. The substance is known as plaster of Paris because the large deposits of Gypsum used for the manufacture of plaster are at Montmeite (Paris).

When plaster of paris is wetted with, it forms a plastic mass which sets in from 5 to 15 minutes to a white porous hard mass. A slight expansion occurs during the setting so that it will take sharp impression of a mould. The process of setting takes place in two steps, the setting step and the hardening step. The final product of setting is gypsum.

$$CaSO_4 \cdot \tfrac{1}{2}H_2O \xrightarrow[\text{setting step}]{3/2\ H_2O}$$

$$CasO_4 \cdot 2H_2O \xrightarrow[\text{Step}]{\text{hardening}} CaSO_4 \cdot 2H_2O$$

The setting step may be catalyzed by NaCl while it is retarded by borax or alum.

Uses: It is used

1. In surgery for plastering the fractured bones.
2. In making moulds for statues, in dentistry etc.
3. In making false ceilings.

SUMMARY

The second group of periodic table is known as alkaline earth metals. Like alkali metals they are reactive. The physical properties and chemical of these elements are explained.

The metallurgy of Mg, its physical and chemical properties are explained in detail. Some compounds of alkaline earth metals such as Epsom salt, calcium sulphate, quick lime, gypsum and plaster of paris are dealt.

REFERENCES:

1. General Chemistry - John Russell McGraw Hill International Editions 3rd Edition.

2. University General Chemistry An Introduction to Chemical Science edited by CNR Rao. McMillan India Limited, Reprint - 2002.

3. Heinemann Advanced Science Chemistry - Second Edition Ann and Patrick Fullick 2000 Heineman Educational Publishers, Oxford.

4. Inorganic Chemistry, P.L. Soni.

CHAPTER – 6

p-BLOCK ELEMENTS

OBJECTIVES

After studying this chapter, you will be able to

* *Understand the nature and properties of p-block elements.*
* *Know the important ores of boron.*
* *Understand the isolation of boron from its ores.*
* *Understand the preparation, properties and uses of boron compounds.*
* *Learn about the allotropes of carbon.*
* *Understand the structure of graphite and diamond and the difference between them.*
* *Acquire knowledge about oxides, carbides, halides and sulphides of carbon group.*
* *Learn about fixation of nitrogen.*
* *Understand the preparation, properties and structure of nitric acid.*
* *Recognize the uses of nitrogen and its compounds.*
* *Know the importance of molecular oxygen and the differences between nascent oxygen and molecular oxygen.*
* *Realise the importance of ozone to life.*

6.1 General Characteristics

The elements belonging to the group 13 to 18 of the periodic table, inwhich p-orbitals are progressively filled, are collectively known as p-block elements.

In all these elements while s-orbitals are completely filled, their p-orbitals are incomplete. These are progressively filled by the addition of one electron as we move from group 13 (ns^2np^1) to group

17 (ns^2np^5). In group 18 (ns^2np^6) both s and p-orbitals are completely filled.

p-block elements show a variety of oxidation state both positive and negative. As we go down the group, two electrons present in the valence `s' orbital become inert and the electrons in the `p' orbital are involved in chemical combination. This is known as `inert pair effect'.

The inert pair effect is really a name, not an explanation. A full explanation involves the decreasing strength of the M-X bond going down the group (for covalent compounds) or the decreasing lattice energies of compounds containing the M^{4+} ion (for ionic compounds). In this way the energy input needed to form compounds of the formula MX_4 are less likely to be balanced by the energy released when the four M-X bonds are formed, so the equilibrium favours the left hand side.

$$MX_2 + X_2 \rightarrow MX_4$$

The existence of a positive oxidation state corresponding to the group number and of another state two units lower is an illustration of the inert pair effect, the term referring to the valence `s' electrons, used in bonding in the higher oxidation state but not in the lower.

With the increase in atomic mass, the ionic character of bonds of the compounds of the group 13(IIIA) elements increases, and some of the heavier metal ions do exist in the +3 oxidation state in aqueous solution. The stability of such compounds with the +3 oxidation state is, however, lower than those with the +1 oxidation state in the case of heavier members of this group. Thus thallium in +1 oxidation state is more stable than in +3 state. This is because, the s electrons in the ns sub-shell do not prefer to form bonds.

This inertness is found only, i) when the `s' electrons are in the fifth or higher principal quantum number ii) when their loss does not afford a species with a noble gas configuration. This property of stabilizing the lower oxidation state keeping the paired electron in the ns orbital is referred to as the `inert pair effect'. This effect is also observed in the elements of groups 12(IIB), 14(IVA) and 15(VA) where the heavier elements exhibit 0, +2 and +3 oxidation states respectively.

Nature of oxides

Oxides of p-block elements may be basic (in case of metallic elements), amphoteric (in case of metalloids) or acidic (in case of non-metals). Non-metals also form a number of oxyacid. In all the groups, the acidic character of the oxide decreases as we move down the group while it increases in the same period from left to right.

For example

Basic oxide	-	Bi_2O_3
Amphoteric oxide	-	SnO, SnO_2, PbO, Pb_2O_3
Acidic oxides	-	SO_3, Cl_2O_7
Oxyacids	-	HNO_3, H_2SO_4.

Basic character increases down the group

CO_2	SiO_2	GeO_2	SnO	PbO
acidic	less acidic	amphoteric	basic	most basic

Acidic character increases across a period

Al_2O_3	SiO_2	P_4O_{10}	SO_2	Cl_2O_7
amphoteric	acidic			most acidic

Nature of hydrides

Many of the p-block elements form hydrides. The hydrides of non-metals are more stable. Thus in any group the stability of the hydride decreases from top to bottom; its strength as an acid also increases in this order. Thus among all the hydrides, hydrogen iodide forms the strongest acid solution in water. In group 15, nitrogen forms the stablest hydride of all. Thus the order of stability of these hydrides is

$$NH_3 > PH_3 > AsH_3 > SbH_3 > BiH_3$$

Nature of halides

Out of the p-block elements, the non-metals form covalent halides. Metallic halides show a gradation from an ionic character to covalent character. As we move from left to right across the period, ionic character of the halides decreases and covalent character increases. For example, $SbCl_2$ is partially ionic whereas $TeCl_4$ is covalent.

In case metals forms halides in more than one oxidation states, halides in lower oxidation state are largely ionic and those in higher oxidation state are largely covalent.

Polarizability of a halide ion depends on its size. Iodides and bromides are more covalent while fluorides are more ionic.

6.2 Group 13 - Boron Group (B, Al, Ga, In, Tl)

Boron does not occur in the free state in nature. In the combined state, it occurs mainly in the form of the salts of boric acid.

6.2.1 Ores of Boron

i)   Boric acid H_3BO_3

ii)   Borax $Na_2B_4O_7.10H_2O$

6.2.2 Extraction

On a large scale, boron is extracted from its minerals, borax $Na_2B_4O_7$ or colemanite $Ca_2B_6O_{11}$. The latter is first converted to borax by boiling with a solution of sodium carbonate in the requisite proportion.

$$2Ca_2B_6O_{11} + 3Na_2CO_3 + H_2O \rightarrow 3Na_2B_4O_7 + 3CaCO_3 + Ca(OH)_2$$

The insoluble calcium carbonate settles down and borax is crystallizedfrom the mother liquor. Boron is isolated from borax in the following two steps.

a) Preparation of boron trioxide:- Borax is treated with hotconcentrated hydrochloric acid, when the sparingly soluble boric acid slowly separates out.

$$Na_2B_4O_7 + 2HCl \rightarrow 2NaCl + H_2B_4O_7$$
$$H_2B_4O_7 + 5H_2O \rightarrow 4H_3BO_3$$

Boric acid is strongly heated when boron trioxide is obtained

$$2H_3BO_3 \rightarrow B_2O_3 + 3H_2O$$

b) Reduction of borontrioxide:- A mixture of borontrioxide withsodium, potassium or magnesium pieces is heated in a crucible to bright redness. The residual boron is broken up and boiled with concentrated HCl to dissolve out magnesium oxide and excess of boric acid when a dark brown powder of amorphous boron is obtained as a residue: It is washed with water and dried.

$$B_2O_3 + 3Mg \rightarrow 2B + 3MgO.$$

Pure boron is obtained in the crystalline form by passing a mixture of boron tribromide vapours and hydrogen over electrically heated filament of tungsten at 1470K. It may also be prepared by submitting a mixture of borontrichloride vapour and hydrogen to the action of a high tension electric arc, when boron is obtained on cooling as a hard black amorphous mass.

Physical properties

Boron exists in two allotropic forms amorphous and crystalline boron. Boron is a non-metallic element and is a non-conductor of electricity.

Chemical properties

1) **Action of air:-** It is unaffected by air at ordinary temperature but whenheated in air to about 975K, it burns forming boron trioxide and a little boron nitride, BN

$$4B + 3O_2 \rightarrow 2B_2O_3$$

$$2B + N_2 \rightarrow 2BN$$

2) **With acids: -** Amorphous boron dissolves in hot concentrated sulphuricand in nitric acid to form boric acid.

$$B + 3HNO_3 \rightarrow H_3BO_3 + 3NO_2$$

$$2B + 3H_2SO_4 \rightarrow 2H_3BO_3 + 3SO_2.$$

3) **With caustic alkali:-** It dissolves in fused caustic alkali and formsboric acid.

4) **As a reducing agent:-** Boron is a powerful reducing agent and caneven replace carbon from carbon dioxide and silicon from silica.

$$3CO_2 + 4B \rightarrow 2B_2O_3 + 3C$$

$$3SiO_2 + 4B \rightarrow 2B_2O_3 + 3Si$$

5) **With metals:-** It combines with metals (except Cu, Ag and Au) at hightemperature in the electric furnace to form borides.

6) **With non-metals:-** Boron combines with nitrogen, chlorine, bromineand carbon at higher temperature forming boron nitride, BN, boron trichloride, BCl_3, boron tribromide, BBr_3 and boron carbide, B_4C respectively. Boron carbide is probably the hardest substance known.

6.2.3 Compounds of Boron

Borax (or) Sodium tetraborate, $Na_2B_4O_7$- Tincal, a crude form ofborax, contains 55% of it and is found in the land dried up lakes of Tibet.

Borax can be prepared

i) **From colemanite:-** It is boiled with concentrated solution of sodiumcarbonate.

$$Ca_2B_6O_{11} + 2Na_2CO_3 \rightarrow 2CaCO_3 + Na_2B_4O_7 + 2NaBO_2.$$

On filtration and concentration, crystals of borax separate. A current of CO_2 is passed through the mother liquor to convert the metaborate into borax.

$$4NaBO_2 + CO_2 \rightarrow Na_2CO_3 + Na_2B_4O_7$$

The residual sodium carbonate is used again for the treatment of a fresh quantity of colemanite.

ii) **From Tincal -** Naturally occurring crude borax (Tincal) is dissolved inwater, filtered, concentrated and crystallized when pure borax is obtained.

Properties

1. When borax is heated above its melting point until all the water of crystallization is expelled, it forms a colourless glassy substance known as borax glass. It then decomposes to give sodium meta borate and boron (III) oxide.

$$Na_2B_4O_7.10H_2O \xrightarrow{\Delta} Na_2B_4O_7 + 10H_2O$$

$$Na_2B_4O_7 \xrightarrow{\Delta} 2NaB + B_2O_3$$

When this mixture is fused with metallic oxide it forms characteristic coloured beads. With the help of the colour, the metal ions can be identified. For example

$$CuO + B_2O_3 \rightarrow Cu(BO_2)_2.$$

Uses: Borax is used

1) to identify the metallic radicals in the qualitative analysis
2) as a flux in welding metals
3) in the manufacture of glass, soap and porcelain
4) as cleaning and dyeing agent in tanneries
5) as a food preservative.

Borax bead test

A pinch of borax is heated in a platinum loop, it melts to give a colourless glassy bead. It is then dipped in a coloured metallic salt solution and again heated. Characteristic coloured beads are formed. From the colour of the beads, the basic radicals are identified. Due to the formation of metallic metaborate, the characteristic colours are formed.

Example: Copper salts give blue beads

In an oxidizing flame

$$CuSO_4 + B_2O_3 \rightarrow Cu(BO_2)_2 + SO_3$$

In a reducing flame

$$2Cu(BO_2)_2 + C \rightarrow 2CuBO_2 + B_2O_3 + CO$$
$$2CuBO_2 + C \rightarrow 2Cu + B_2O_3 + CO$$

Borax bead test is used to identify the coloured salts.

Metallic compounds	Colour in oxidizing Flame	Colour in reducing flame
Copper	Blue	Red
Iron	Yellow	Bottle green
Manganese	Pinkish violet	Colourless
Cobalt	Blue	Blue
Chromium	Green	Green
Nickel	Brown	Grey

6.3 Carbon group elements

The elements carbon, silicon, germanium, tin and lead constitute the 14th group of the periodic table. These are p-block elements having the configuration ns^2np^2.

Element	At.No.	Electronic structure
Carbon	6	[He] $2s^2\,2p^2$
Silicon	14	[Ne] $3s^2\,3p^2$
Germanium	32	[Ar] $3d^{10}\,4s^2\,4p^2$
Tin	50	[Kr] $4d^{10}\,5s^2\,5p^2$
Lead	82	[Xe] $4f^{14}\,5d^{10}\,6s^2\,6p^2$

6.3.1 Allotropic forms of carbon

Carbon exhibits allotropy and occurs as

i)   Diamond, a beautiful crystalline substance
ii) Graphite, a soft greyish black crystalline substance
iii) Amorphous carbon, black residue left when carbon compounds are heated.

Different amorphous varieties of Carbon are (i) Coal, (ii) Coke, (iii) Charcoal, (iv) Bone black or Animal charcoal, (v) lampblack, (vi) carbon black, (viii) Gas carbon and (ix) petroleum coke.

6.3.2 Structure of diamond

In diamond every atom is bonded with the other by covalent links resulting in the formation of giant molecule. Each carbon atom is linked with four neighboring carbon atoms held at the corners of a regular tetrahedron by covalent bonds. The C-C bonds are very strong. The crystal of diamond is very hard and has high melting and boiling points.

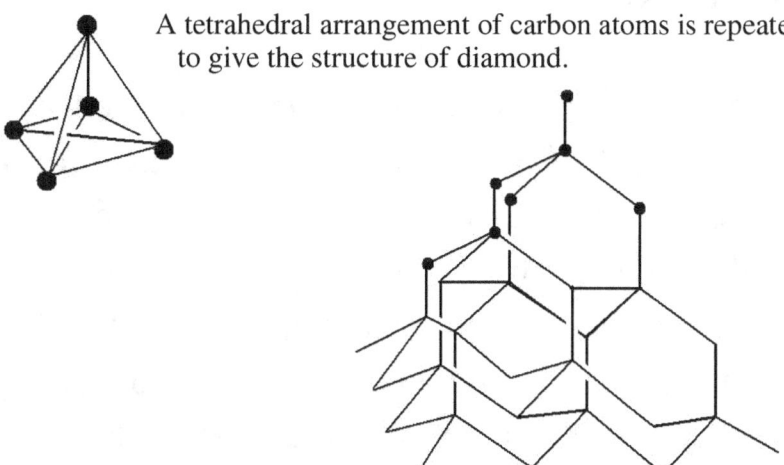

A tetrahedral arrangement of carbon atoms is repeated to give the structure of diamond.

Structure 7.3:

The combined strength of the many carbon-carbon bonds within the structure of diamond give it both great hardness and a lack of chemical reactivity.

Structure of graphite

It consists of separate layers. The carbon atoms are arranged in regular hexagons in flat parallel layers. There is no strong bonding between different layers, which are, therefore, easily separable from each other. Since there are no covalent linkages between the adjacent planes, graphite can be easily cleaves along the lines of the planes. Whilst the bonds within the layers are strong, those between the layers are not and so they slide over each other easily This accounts for the softness and lubricating power of graphite.

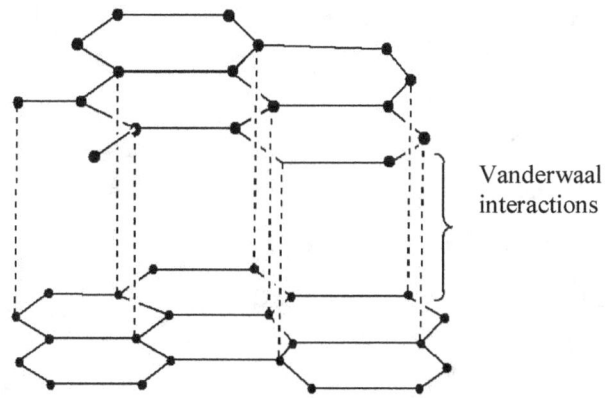

Vanderwaal interactions

Structure 7.4: The structure of graphite.

Structure of Buckminster fullerenes
Fullerenes

In 1985, a new allotrope of carbon was discovered by Richard Smalley and Robert Curl of Rice University, Texas, working with Harry Kroto of Sussex University. The first to be identified and the most symmetrical of the family, with 60 atoms and 32 sides (20 hexagons and 12 pentagons), was nick named `buckyball' and was then named buck minister fullerene, because it resembles the geodesic domes developed by an American inventor called R.Buckminister fuller. The group of spherical carbon molecules is called fullerenes. These compounds have superconducting properties and its potential for opening new areas of chemistry have madestudy of the `buckyball' as one of the most rapidly expanding areas of chemical research.

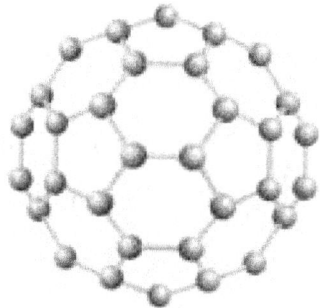

Fullerene

Amorphous form of carbon

Amorphous carbon is the most reactive form of carbon. It burns relatively easily in air, thereby serving as a fuel, and is attacked by strong oxidizing agents. This form has structural features of graphite, such as sheets and layers. It's atomic structure is much more irregular.

General properties
6.3.4. Metallic character

Carbon and silicon are non-metals, germanium is a metalloid while tin and lead are metals. Thus metallic character increases on descending the group since ionization energy decreases on descending the group.

Hydrides

All of these elements form covalent hydrides though the number of hydrides and the ease with which these are formed decreases from carbon to lead. Carbon gives a vast number of hydrides (alkanes), silicon and germanium (silanes and germanes) whereas stannane (SnH_4) and plumbane (PbH_4) are the only hydrides of tin and lead are known.

Unlike alkanes, silanes are strong reducing agents, explode in chlorine and are readily hydrolyzed by alkaline solutions. The difference is probably due to the difference in electronegativity between C and Si resulting in difference between C-H and Si-H linkages.

Halides

All these elements give tetra halides. Tetrachlorides are usually fuming liquids at ordinary temperature. Carbon tetrahalide resists hydrolysis. This is because due to the absence of d-orbitals. Maximum covalency of carbon is only four and there is no possibility of formation of coordinate linkages with H_2O, which could lead to hydrolysis.

Tetrahalides of rest of the elements undergo hydrolysis. For example

$$SiX_4 + 2H_2O \rightarrow SiO_2 + 4HX$$

Carbon, silicon and germanium form trihalides of the type MHX_3. Lead and tin do not form trihalides. Silicon, germanium, tin and lead form dihalides.

Chlorides

1. The chlorides are all simple molecular substances with tetrahedral molecules.
2. The stability of the chlorides decreases down the group and the +2 oxidation state becomes more stable than the +4 state. Only tin and lead form chlorides in which their oxidation state is +2, the other chlorides existing solely in the +4 state. Tin(II) chloride is a solid that is soluble in water, giving a solution which conducts electricity. It is also soluble in organic solvents. Its melting point is 246°C. Lead(II) chloride is also a solid. It is

sparingly soluble in water. The chlorides of the group 14 elements in their +4 oxidation state illustrate further the change in character of the elements from non-metal to metal down the group and giving a solution which conducts electricity, and melts at 501°C. These observations suggest that tin(II) chloride has both covalent and ionic character, while lead(II) chloride is predominantly ionic.

3. All the chlorides with +4 oxidation state are readily hydrolyzed by water, except tetrachloromethane (CCl_4).

Carbides

Compounds of carbon with less electronegative elements (eg. metals, Be, B, Si etc.) are called carbides. These are of three main types.

i) Ionic or salt-like eg. acetylides, methanides, allylides

ii) Interstitial or metallic eg. WC and

iii) Covalent eg. B_4C, SiC.

All the three types of carbides are prepared by heating the element or its oxide with carbon or a hydrocarbon to a high temperature.

$$2Be + C \rightarrow Be_2C$$
$$CaO + 3C \rightarrow CaC_2 + CO$$
$$SiO_2 + 3C \rightarrow SiC + 2CO$$

Oxides

1. The oxides show a marked trend in structure from the molecules of carbondioxide to giant structures intermediate between ionic and covalent lower down the group.

2. The +2 oxidation state is the more stable state in the case of leadoxide, and lead(IV) oxide decomposes on heating giving lead(II) oxide, a solid that melts at 886°C. The structure of lead(II) oxide is predominantly ionic.

3. The oxides at the top of the group (CO_2 and SiO_2) have an acidic nature, the carbonate ion CO_3^{2-} being produced easily in dilute aqueous solutions. The ease of formation of Oxo anions (SiO_3^{2-},

GeO_3^{2-} etc.) decreases down the group as the acidic character decreases. The oxides of germanium, tin and lead are amphoteric, reacting to form simple salts with acids.

Uses of carbon and its compounds

1. Carbon and its compounds play an enormous role in the global economy, eg. Fossil fuels.
2. Halogenated carbon compounds are used as refrigerants, aerosol propellants, fire extinguisher and solvents.
3. CS_2 is used in the manufacture of viscose rayon (artificial silk) and cellophane.

6.4 Nitrogen Group

The elements nitrogen, phosphorus, arsenic, antimony and bismuth constitute 15th group of the periodic table. This group is called nitrogen group. These elements have the general electronic configuration ns^2np^3. Allthese elements have five electrons in their outermost orbitals. The `s' orbital contains two electrons and p orbital contains three electrons. These three electrons are equally distributed in three p-orbitals as p_x^1 p_y^1 p_z^1 which correspond to half-filled configuration.

As we go down the group, the two electrons present in the valence `s' orbital become inert and only the three electrons present in the outermost p-orbitals are involved in chemical combinations. This is known as inert pair effect. As we move from nitrogen to bismuth, the pentavalency becomes less pronounced while trivalency becomes more pronounced.

Nitrogen was discovered in 1772 by Daniel Rutherford, a Scottish physician and chemist. Elementary nitrogen constitutes three-fourths of air by weight. It is also abundant in the combined state as saltpeter (KNO_3), sodium nitrate (chile saltpeter) and ammonium salts. Nitrogen is an essential constituent of all vegetable and animal proteins.

Fixation of nitrogen

The nitrogen present in the atmosphere is free or elementary nitrogen whereas nitrogen present in various nitrogenous compounds is called combined or fixed nitrogen. The conversion of free atmospheric nitrogen to a nitrogen compound is called fixation of nitrogen.

Method employed for fixation or bringing atmospheric nitrogen into combination:

Manufacture of ammonia (Haber's process):

A mixture of nitrogen and hydrogen in the ratio 1:3 under pressure (200-900 atm) is passed over a catalyst finely divided iron and molybdenum as promoter, heated to about 770K.

$$N_2 + 3H_2 \rightleftharpoons 2NH_3$$

The ammonia so manufactured can be oxidized to nitric oxide by passing a mixture of ammonia and air over heated platinum gauze at 1070K. Nitric oxide combines with more of oxygen to give nitrogen dioxide which when absorbed in water in the presence of excess of air, gives nitric acid (Ostwald's process)

$$4NH_3 + 5O_2 \rightarrow 4NO + 6H_2O$$
$$2NO + O_2 \rightarrow 2NO_2$$
$$4NO_2 + 2H_2O + O_2 \rightarrow 4HNO_3$$

Ammonia and nitric acid manufactured above may be converted into ammonium salts and nitrates suitable as fertilizers. Thus these methods of nitrogen fixation are of vital importance to the agriculturists.

Nitrogen fixation in nature

Due to electrical disturbances atmospheric nitrogen and oxygen combine to give nitric oxide which gets further oxidized to nitrogen dioxide. This reacts with rain water in the presence of excess of oxygen to produce nitric acid and is washed down to earth. Here it reacts with bases of the soil to give nitrates.

In addition to this, certain bacteria living in the nodules on roots of leguminous plants e.g. pea, beans etc., convert nitrogen into nitrogenous compounds which can be directly assimilated by the plant.

Basics of Inorganic Chemistry

Nitrogen cycle

There is a continual turnover of nitrogen between the atmosphere, the soil, the sea and living organisms. The nitrogen passes from atmosphere to plants and animals, converted into useful products like ammonia, nitric acid etc. and still its percentage in the atmosphere remains practically unchanged. This is due to the fact that combined nitrogen is constantly passing back to the atmosphere. This cycle of changes involved is known as nitrogen cycle.

Uses of nitrogen compounds

1. Liquid ammonia is used as solvent.
2. Ammonia is used as a refrigerant in ice-plants.
3. Ammonia is used in the manufacture of artificial silk, urea, manures, washing soda etc.
4. Nitrous oxide mixed with oxygen is used as anesthetics for minor operations in dentistry and surgery.
5. Nitrous acid is used in the manufacture of azo-dyes.
6. Nitric acid is used in the manufacture of fertilizers, explosives like TNT, GTN, etc.
7. Nitric acid is used in the purification of gold and silver.
8. Nitric acid is used in pickling of stainless steel.
9. Nitric acid is used in the manufacture of perfumes, artificial silk, medicines etc.
10. Liquid nitrogen is used as a refrigerant.

6.4.2 Nitric acid

Nitric acid is an important oxyacid of nitrogen. It was called as `aqua tortis' by alchemists. It means strong water. It was first prepared by Glauber (1650). Later Cavendish (1784) stated that nitric acid may be formed by passing electric sparks through the mixture of nitrogen and moist oxygen. Traces of nitric acid occur in air where it is formed by electric sparks through the mixture of nitrogen and moist oxygen. Traces of nitric acid occur in air where it is formed by electric discharges and is washed down by rain.

Preparation

1. Laboratory preparation

Nitric acid is prepared in the laboratory by heating a nitrate salt with concentrated sulphuric acid.

$$NaNO_3 + H_2SO_4 \rightarrow NaHSO_4 + HNO_3$$

Vapours of nitric acid are condensed to a brown liquid in a receiver cooled under cold water. Dissolved oxides of nitrogen are removed by redistillation or blowing a current of carbondioxide or dry air through the warm acid.

2. Manufacture of nitric acid

1. Nitric acid is manufactured by blowing air into an electric arc struck between two water cooled copper electrodes and spread into a disc with the help of a magnetic field at right angle. The serious disadvantage of the method is now obsolete.

2. **Ostwald's process**

   Large quantities of ammonia manufactured by Haber's process are converted into nitric acid by Ostwald's process.

$$4\,NH_3 \xrightarrow[1155K]{Platinum\ gauze} 4NO + 6H_2O$$

$$2NO + O_2 \longrightarrow 2NO_2$$

$$4NO_2 + 2H_2O + O_2 \longrightarrow 4HNO_3$$

Dilute nitric acid may be concentrated by distillation until a constant boiling point mixture is obtained (98%). Fuming nitric acid is obtained by distilling this acid with concentrated sulphuric acid. Crystals of pure nitric acid may be obtained by cooling 98% acid in a freezing mixture.

Properties

Physical properties

1. It is a colourless fuming liquid when pure, but may be coloured yellow by its dissociation products mainly nitrogen dioxide.

2. It has extremely corrosive action on the skin and causes painful sores.

3. Pure acid has a specific gravity of 1.54. It boils at 359K and freezes to a white solid (m.p. 231K).

6.5 Oxygen group - group 16

The elements oxygen, sulphur, selenium, tellurium and polonium constitute 16th group of the periodic table. The first four elements are non metals. Collectively they are called the 'chalcogens' or ore - forming elements. This is because a large number of metals are oxides or sulphides.

Oxygen is a very important element in inorganic chemistry, since it reacts with almost all the other elements. Oxygen is the most abundant of all elements. It exists in the free form as dioxygen or molecular oxygen and makes up 20.9% by volume and 23% by weight of the atmosphere.

6.6 Importance of molecular oxygen

Hemoglobin is an iron containing coordination compound in red blood cells responsible for the transport of oxygen from the lungs to various parts of the body. Myoglobin is a similar substance in muscle tissue, acting as a reservoir for the storage of oxygen and as a transport of oxygen within muscle cells.

Hemoglobin consists of heme, a complex of Fe(II) bonded to a porphyrin ligand and globin protein. The sixth position is vacant in free hemoglobin but is occupied by oxygen in oxyhaemoglobin. Hemoglobin (Hb) and O_2 are in equilibrium with oxyhaemoglobin.

$$Hb + O_2 \rightleftharpoons \underset{\text{Oxyhaemoglobin}}{HbO_2}$$

Oxyhaemoglobin is formed in the lungs and carried to the cells, where it gives up its oxygen.

Hemoglobin then binds with HCO_3^- which is formed by the reaction of CO_2 (released by the cell) with water. After reaching the lungs, due to hydrolysis CO_2 is released.

Most of the O_2 has been produced by photosynthesis. The dioxygen (or) molecular oxygen is prepared by the green plants. The chlorophyll in the green parts of the plants uses the solar energy to make carbohydrate and molecular oxygen. Oxygen makes up 46.6% by weight of the earth's crust, where it is the major constituent of silicate minerals.

Practically all the elements react with dioxygen either at room temperature (or) on heating except Pt, Au, W and Noble gases. Eventhough the bond energy of oxygen is high (493 kJ mol^{-1}), the reactions are generally strongly exothermic and once started often continue spontaneously.

Dioxygen is also called as molecular oxygen. The molecular oxygen is essential for respiration (for the release of energy in the body) by both animals and plants. It is therefore essential for life. Hence molecular oxygen acts as a cell fuel.

The complex formed between dioxygen and hemoglobin (the red pigment in blood) is of vital importance. Since it is the method by which higher animals transport dioxygen around the body to the cells.

6.6.1 Nascent oxygen and molecular oxygen

Oxygen molecule is very stable. It dissociates only to a small extent when heated to a very high temperature. This reaction is endothermic

$$O_2 \rightarrow\!\!> 2@ \ \hat{u}+ \qquad NFDO$$

However, when an electric discharge is passed through oxygen at a very low pressure, it dissociates to the extent of about 20%.

For example when oxygen is passed at about 1 mm pressure through a discharge tube, the resulting gas is found to be chemically more reactive. Its line spectrum shows that it consists of the free atoms. Hence atomic oxygen is formed according the following endothermic reactions.

$$O_2 \ \rightarrow 2 \ 2 \ \hat{u}+ \qquad N\text{-}$$

Reactions

1. Formation of molecular oxygen

When a thin platinum wire is placed in atomic oxygen, it quickly gets heated up and begins to glow due to the recombination of oxygen atoms with liberation of heat energy. The rise of temperature of platinum wire under standardized conditions is a measure of the concentration of the atomic oxygen in the gas.

2. Formation of ozone

Atomic oxygen combines with molecular oxygen to give ozone which may be condensed by means of liquid air

$$O_2 + [O] \longrightarrow O_3$$

3. Oxidation

Atomic oxygen is an extremely powerful oxidizing agent and oxidizes aliphatic and aromatic hydrocarbons and methyl alcohol with emission of heat and light. With nitric oxide, a characteristic greenish - white luminescence is produced. H_2S and CS_2 react with it and burst into greyish blue coloured flame.

6.6.2 Oxides

Generally all the elements react with dioxygen to form oxides. Oxides are binary compounds of oxygen. Oxides may be classified depending on their structure (or) their chemical properties.

i) Acidic oxides

The oxides of non-metals are usually covalent and acidic. They have low melting and boiling points, though some B_2O_3 and SiO_2 form infinite "giant molecules" and have high melting points. They are all acidic. Some oxides dissolve in water and thus forming acids. Hence they are called as acid anhydrides

$$B_2O_3 + 3H_2O \rightarrow 2H_3BO_3$$
$$N_2O_5 + H_2O \rightarrow 2HNO_3$$
$$P_4O_{10} + 6H_2O \rightarrow 4H_3PO_4$$
$$SO_3 + H_2O \rightarrow H_2SO_4$$

others which do not react with water such as SiO_2 reacts with NaOH and shows acidic properties.

ii) Basic oxides

Metallic oxides are generally basic. Most metal oxides are ionic and contain the O^{2-} ion. Some oxides dissolve in water and form alkaline solution.

$Na_2O + H_2O \rightarrow 2NaOH$

$BaO + H_2O \rightarrow Ba(OH)_2$

Many metal oxides with formula M_2O_3 and MO_2, though ionic, do not react with water.

Examples: Tl_2O_3, Bi_2O_3 and ThO_2.

These oxides react with water to form salts and hence they are bases.

$$CaO + 2HCl \rightarrow CaCl_2 + H_2O$$

If a metal exists in more than one oxidation state and they form more than one oxide

$$eg. \; CrO, \; Cr_2O_3, \; CrO_3, \; PbO, \; PbO_2$$

iii) Amphoteric oxides

The oxides which react with both strong acids and strong bases are called as amphoteric oxides.

$ZnO + 2NaOH \rightarrow Na_2ZnO_2 + H_2O$

Sodium zincate

$ZnO + 2HCl \rightarrow ZnCl_2 + H_2O$

iv) Peroxides

These oxides contain more oxygen than would be expelled from the oxidation number of M. Some are ionic and contains the peroxide ion O_2^{2-}. The metal belonging to the group I and II (Na_2O_2, BaO_2) contain O_2^{2-} ion. Others are covalently bound and contain -O-O- in the structure.

Oxides such as PbO_2 react with acids liberate Cl_2

$PbO_2 + 4HCl \rightarrow PbCl_2 + 2H_2O + Cl_2$

v) Compound oxides

Some oxides behave as if they are compounds of the two oxides. Ex. Ferrous-ferric oxide(Fe_3O_4). This is considered to be the mixture of FeO and Fe_2O_3.

They react with acids and forms a mixture of ferrous and ferric salts.

$$Fe_3O_4 + 8HCl \rightarrow FeCl_2 + 2FeCl_3 + 4H_2O$$

vi) Neutral oxides

A few covalent oxides have no acidic (or) basic properties (N_2O, NO, CO).

vii) Dioxides

They also contain higher proportion of O_2 than expected. But they do not liberate H_2O_2 with acid.

Ex. NO_2, SO_2

6.7 Ozone

Ozone is an allotropic form of oxygen and its molecular formula is O_3. It is an unstable dark blue diamagnetic gas. The presence of ozone in extremely small quantities has been observed in the atmosphere in places near the seaside (or) big lakes. It is present in sufficient quantities in the atmosphere at attitudes of 12 to 15 miles above the earth's surface. Ozone is particularly important since there is a layer of ozone in the upper atmosphere which absorbs harmful UV radiations from the sun and protects the people and other living organisms on the earth.

Laboratory preparation

Ozone is prepared in the laboratory by passing silent electrical discharges through dry oxygen in an apparatus known as the ozoniser. The commonly used ozoniser is Siemen's ozoniser

(i) Siemen's ozoniser

It consists of two concentric metal tubes sealed together at one end. The inner side of the inner tube and the outer side of the outer tube are coated with tin foil and connected to one terminal each of an induction coil. A current of pure dry oxygen at low temperature is passed through annular space between the two tubes and by the silent action of electric discharge, the oxygen is partially converted into ozone. The sample of gas escaping from ozoniser is called ozonized oxygen and contains about 12% ozone.

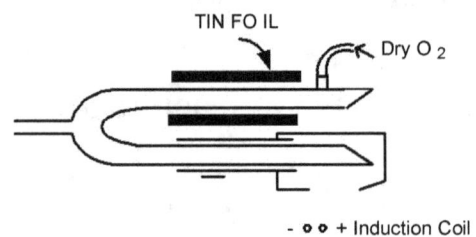

Properties
(i) Physical properties

It is a light blue gas which condenses at 160.6 K into a dark blue liquid. This liquid freezes at 23.3 K.

Chemical properties

1) Decomposition: Pure ozone decomposes with an explosive violence.

$$2O_3 \rightarrow 3O_2$$

2) Oxidizing action: Since it can liberate an atom of nascent oxygen easily ($O_3 \rightarrow O_2 + O$) it acts as a powerful oxidizing agent.

i) Lead sulphide is oxidized to lead sulphate

$$PbS + 4O_3 \rightarrow PbSO_4 + 4O_2$$

ii) Potassium manganate is oxidized to potassium permanganate

$$2K_2MnO_4 + H_2O + O_3 \rightarrow 2KMnO_4 + 2KOH + O_2$$

3) Ozone reacts with peroxides and reduces it to oxides with the liberation of oxygen.

$$BaO_2 + O_3 \rightarrow BaO + 2O_2$$
$$H_2O_2 + O_3 \rightarrow H_2O + 2O_2$$

Uses of ozone

1) It is used as germicide and disinfectant.
2) It is used for bleaching oils, ivory, flour, starch, etc.
3) Used in the manufacture of artificial silk and synthetic camphor.

Ozone structure

The ozone structure molecule consists of three oxygen atoms having a bent

Each O atom contributes six valence electrons and so the total

3x6=18 electrons.

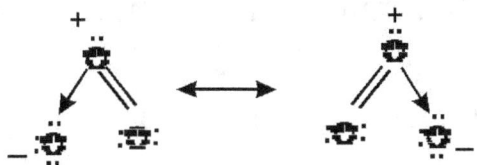

Ozone molecule is said to be resonance hybrid of the two contributing forms I & II.

Ozone layer

Ozone is produced in the upper atmosphere through absorption of aSKRWRQ K# RI XOWUDYLROHW OLJKW E\ DQ 2_2 molecule.

$$O_{2(g)}K\# \rightarrow 2O_{(g)}$$
$$O_{(g)} + O_{2(g)} \rightarrow O_{3(g)}$$

The ozone molecule formed has an excess of energy and dissociates back to O_2 and O and it reacts with another molecule (M) such as CO_2, N_2 or O_2, which causes the excess energy thus stabilizes the ozone molecule

$$O^*_{3(g)} + M_{(g)} \rightarrow O_{3(g)} + M^*_{(g)}$$

Factors affecting ozone layer

The ozone in the upper atmosphere is important in shielding us from the intense ultraviolet radiation coming from the sun. The so-called ozone shield is a shell about 30 km altitude which contains enough ozone to absorb short wavelength UV radiation (less than 300 nm). Hence ozone is considered to be 'earth's protective umbrella'. The absorption causes dissociation of O_3 to reform O_2.

$$O_{3(g)}K^\# \rightarrow O_{2(g)} + O_{(g)}$$
$$2O_{(g)} \rightarrow O^\bullet_{2(g)}$$
$$O^\bullet_{2(g)} + M_{(g)} \rightarrow O_{2(g)} + M^\bullet_{(g)}$$

Existence of ozone shield owes to the life on the earth, since living tissues are very sensitive to wavelengths of ultraviolet absorbed by ozone. In recent years, the shield is damaged mainly by supersonic aircraft andchlorofluorocarbon products in the jet exhaust reduce ozone, and decrease its concentration in the shield.

Chlorofluorocarbon reacts with O_3 and causes a hole in the ozone layer. CFC's are used as refrigerants and as propellants in some "aerosol sprays". The lifetimes of CFCs are so long that in another decades, the extent of ozone depletion in the upper atmosphere will be tremendous. It is reported that the holes caused in the ozone layer over the Antarctic and Arctic ocean are due to the use of CFCs in aerosols and refrigerators. It is feared that this will allow an excessive amount of UV light to reach the earth which will cause skin cancer (melanoma) in human.

UV

$$CFC \rightarrow Cl_{(g)}$$
$$Cl_{(g)} + O_{3(g)} \rightarrow ClO_{(g)} + O_{2(g)}$$
$$ClO + O_{(g)} \rightarrow Cl_{(g)} + O_{2(g)}$$
$$O_{3(g)} + O_{(g)} \rightarrow 2O_{2(g)}$$

It is also seen that the oxides of nitrogen (from car exhausts) and the halogen can damage the ozone layer. Therefore the protecting shield of the earth must be protected by taking immediate steps over the control of pollution.

SUMMARY

Groups 13 to 18 of the periodic table are known as p-block elements. The lower oxidation states of these elements are stabilized by inert pair effect.

Group 13 is known as Boron group. The element boron is extracted from its ore borax and colemanite. It reacts to give many compounds. Most important of them is borax, which is used to identify the metallic radicals in the qualitative analysis.

Group 14 is known as carbon group. Carbon exists in different allotropic forms such as diamond, graphite, fullerenes and other amorphous form. The elements of this group form various hydrides, oxides, halides and carbides.

Group 15 is known as nitrogen group. The element nitrogen is essential for plant life. It plays a vital role in fixation of nitrogen and the importance can be studied by the nitrogen cycle.

Nitric acid is the important oxyacid of nitrogen. It is prepared by Ostwald process. It oxidizes metals, non-metals, compounds, etc.

Group 16 is known as oxygen group. Oxygen is the essential element for life. Dioxygen or molecular oxygen plays an important role in functioning of hemoglobin and myoglobin. The study about ozone and the depletion of ozone layer in the upper atmosphere is very essential. The causes of ozone depletion must be considered seriously and steps should be taken to stop the depletion.

REFERENCES:

1. General Chemistry - John Russell McGraw Hill International Editions 3rd Edition.
2. University General Chemistry An Introduction to Chemical Science edited by CNR Rao. McMillan India Limited, Reprint - 2002.
3. Heinemann Advanced Science Chemistry - Second Edition Ann and Patrick Fullick 2000 Heineman Educational Publishers, Oxford.
4. Inorganic Chemistry, P.L. Soni.